INTERNATIONAL SERIES OF MONOGRAPHS IN
NATURAL PHILOSOPHY

GENERAL EDITOR: D. TER HAAR

VOLUME 34

ELEMENTS OF HAMILTONIAN MECHANICS

OTHER TITLES IN THE SERIES OF NATURAL PHILOSOPHY

Vol. 1. DAVYDOV—Quantum Mechanics
Vol. 2. FOKKER—Time and Space, Weight and Inertia
Vol. 3. KAPLAN—Interstellar Gas Dynamics
Vol. 4. ABRIKOSOV, GOR'KOV and DZYALOSHINSKII—Quantum Field Theoretical Methods in Statistical Physics
Vol. 5. OKUN'—Weak Interaction of Elementary Particles
Vol. 6. SHKLOVSKII—Physics of the Solar Corona
Vol. 7. AKHIEZER et al—Collective Oscillations in a Plasma
Vol. 8. KIRZHNITS—Field Theoretical Methods in Many-body Systems
Vol. 9. KLIMONTOVICH—Statistical Theory of Non-equilibrium Processes in a Plasma
Vol. 10. KURTH—Introduction to Stellar Statistics
Vol. 11. CHALMERS—Atmospheric Electricity (2nd Edition)
Vol. 12. RENNER—Current Algebras and their Applications
Vol. 13. FAIN and KHANIN—Quantum Electronics, Volume 1—Basic Theory
Vol. 14. FAIN and KHANIN—Quantum Electronics, Volume 2—Maser Amplifiers and Oscillators
Vol. 15. MARCH—Liquid Metals
Vol. 16. HORI—Spectral Properties of Disordered Chains and Lattices
Vol. 17. SAINT JAMES, THOMAS, and SARMA—Type II Superconductivity
Vol. 18. MARGENAU and KESTNER—Theory of Intermolecular Forces
Vol. 19. JANCEL—Foundations of Classical and Quantum Statistical Mechanics
Vol. 20. TAKAHASHI—Introduction to Field Quantization
Vol. 21. YVON—Correlations and Entropy in Classical Statistical Mechanics
Vol. 22. PENROSE—Foundations of Statistical Mechanics
Vol. 23. VISCONTI—Quantum Field Theory, Volume 1
Vol. 24. FORTH—Fundamental Principles of Modern Theoretical Physics
Vol. 25. ZHELEZNYAKOV—Radioemission of the Sun and Planets
Vol. 26. GRINDLAY—An Introduction to the Phenomenological Theory of Ferroelectricity
Vol. 27. UNGER—Introduction to Quantum Electronics
Vol. 28. KOGA—Introduction to Kinetic Theory. Stochastic Processes in Gaseous Systems
Vol. 29. GALASIEWICZ—Superconductivity and Quantum Fluids
Vol. 30. CONSTANTINESCU and MAGYARI—Problems in Quantum Mechanics
Vol. 31. KOTKIN and SERBO—Collection of Problems in Classical Mechanics
Vol. 32. PANCHEV—Random Functions and Turbulence
Vol. 33. TALPE—Theory of Experiments in Paramagnetic Resonance

ELEMENTS OF HAMILTONIAN MECHANICS

SECOND EDITION

BY

D. TER HAAR

University Reader in Theoretical Physics
and
Fellow of Magdalen College, Oxford

PERGAMON PRESS

OXFORD · NEW YORK · TORONTO
SYDNEY · PARIS · FRANKFURT

U.K.	Pergamon Press Ltd., Headington Hill Hall, Oxford OX3 0BW, England
U.S.A.	Pergamon Press Inc., Maxwell House, Fairview Park, Elmsford, New York 10523, U.S.A.
CANADA	Pergamon Press Canada Ltd., Suite 104, 150 Consumers Rd., Willowdale, Ontario M2J 1P9, Canada
AUSTRALIA	Pergamon Press (Aust.) Pty. Ltd., P.O. Box 544, Potts Point, N.S.W. 2011, Australia
FRANCE	Pergamon Press SARL, 24 rue des Ecoles, 75240 Paris, Cedex 05, France
FEDERAL REPUBLIC OF GERMANY	Pergamon Press GmbH, 6242 Kronberg-Taunus, Hammerweg 6, Federal Republic of Germany

First edition 1961 and
Second edition 1964 published by
North-Holland Publishing Co., Amsterdam

First Pergamon edition 1971
Reprinted 1982

Library of Congress Catalog Card No: 70-150691

The present edition is an unaltered photographic reproduction of the 1964 edition.

Printed in Great Britain by A. Wheaton & Co. Ltd., Exeter

ISBN 0-08-016726-8

PREFACE

For the last five years I have been giving a course of lectures on classical mechanics to last year undergraduates and first year research students at Oxford University. This course consisted of about 30 lectures and covered the subject matter of the present textbook. I devoted to each of the chapters roughly the same amount of time, namely about four hours, and feel that this particular course adequately covered those aspects of classical mechanics with which any physicist should be acquainted these days, without really delving into the many beautiful ramifications of which there are so many in classical mechanics. As the emphasis and the selection of subject matter in my course is rather different from the one in existing modern textbooks, of which those by Corben and Stehle and by Goldstein are the best known and most widely used examples, I felt that it might be of use to some people to publish this volume.

In writing this text I have benefited greatly from my own notes on lectures given by the late Professor H. A. Kramers at Leiden. I am also in debt to the various Oxford undergraduates whom I have taught classical mechanics. Finally, I should like to express my gratitude to Professor J. de Boer, Professor W. E. Lamb Jr., and Dr. W. E. Parry for critically reading through the manuscript of this volume and suggesting possible improvements.

D. ter Haar

Oxford, December 1960

PREFACE TO THE SECOND EDITION

Apart from a few corrections of misprints in the text, the only difference between this edition and the first one lies in the inclusion of a larger number of problems, most of which have been taken from Oxford University examination papers. I should like to express my gratitude to the Oxford University Press for permission to include these problems.

Oxford, June 1963 D. ter Haar

CONTENTS

CHAPTER 1

NEWTONIAN MECHANICS

In this chapter we discuss briefly Newton's laws and apply them to the case of a central field of force with special reference to the inverse square law force. Scattering by a central field of force is briefly considered. Some properties of systems consisting of several particles which interact through two-body central forces are also discussed.

1. NEWTON'S LAWS

The basis of classical mechanics is formed by Newton's laws with which we shall start our discussion. We shall assume that the terms used in stating these laws have a well-defined meaning (they have all an intuitive meaning), as we do not wish to discuss the various concepts introduced here. In writing down Newton's laws we are assuming that there are systems of reference in which they are valid. Such systems are called *inertial systems*, and we shall assume that all our vectors are defined in such a system. We may add that any system moving with a uniform velocity with respect to an inertial system is again an inertial system.

We can now state Newton's laws:

Lex prima. If there are no forces acting on a particle it will persist in its motion, that is, it will move along a straight line with a constant velocity.

Lex secunda. If there are forces acting on a particle, the rate of change of the linear momentum of the particle will be equal to the total force acting on it. The linear momentum of a particle is defined as the product of its mass and its velocity.

Lex tertia. When two particles act upon one another, the force due to the first particle upon the second one is equal to, but in the opposite direction to, the force due to the second particle upon the first one. (*Actio est reactio*).

In using the term *particle* we shall throughout the book have in mind a point particle, that is, an entity characterised by its mass m, its position $\boldsymbol{x}$ and its velocity $\boldsymbol{v}$ which is equal to the rate of change of $\boldsymbol{x}$,

$$\boldsymbol{v} = \frac{\mathrm{d}\boldsymbol{x}}{\mathrm{d}t}, \tag{1.101}$$

where t is the time coordinate.

As long as we do not discuss relativistic effects, we can assume m to be a constant which is characteristic of the particle.

We can express Newton's laws in mathematical form as follows:

Lex prima: If $\boldsymbol{F} = 0, \quad \boldsymbol{v} = \text{constant};$ (1.102)

Lex secunda: $\frac{\mathrm{d}}{\mathrm{d}t} m\boldsymbol{v} = \boldsymbol{F};$ (1.103)

Lex tertia: $\boldsymbol{F}_{12} = -\boldsymbol{F}_{21}$, or, $\boldsymbol{F}_{12}+\boldsymbol{F}_{21} = 0.$ (1.104)

In these equations $\boldsymbol{F}$ is the total force on a particle, and $\boldsymbol{F}_{12}$ ($\boldsymbol{F}_{21}$) is the force exerted by particle number 2 (1) upon particle number 1 (2).

If m is constant, equation (1.103) can also be written in the form

$$m\boldsymbol{a} = \boldsymbol{F}, \qquad \boldsymbol{a} = \frac{\mathrm{d}\boldsymbol{v}}{\mathrm{d}t}, \tag{1.105}$$

where $\boldsymbol{a}$ is the acceleration of the particle. This form of Newton's second law – force equals mass times acceleration — is the slightly more familiar one, but it is interesting to note that Newton himself used the other formulation which remains valid in the case of variable m.

The first law is Galileo's inertial law; hence the term inertial system. Comparing (1.105) and (1.102) we see that the first law is a special case (for $\boldsymbol{F} = 0$) of the second law. The mass m which can be considered to be defined by (1.103) is called the *inertial mass* of the particle; it has experimentally been shewn to be equal to the *heavy* (or gravitational) *mass* of a particle which is proportional to its weight. It may be mentioned here that this equality of the two kinds of mass follows naturally in the general theory of relativity.

Before discussing some of the consequences of Newton's laws we wish to mention that one sometimes finds as a lex quarta the rule that the addition of forces acting on a point particle follows the rules for the addition of vectors. This has tacitly been assumed in our equations (1.103) and (1.104), as we have used for the force a letter type typical of vectors.

Even without specifying the forces, we can draw some conclusions from (1.103) and (1.104). First of all, let us consider a system of two particles where the only forces acting upon the particles are $\boldsymbol{F}_{12}$ and $\boldsymbol{F}_{21}$. From (1.104) it follows that

$$\int_{t'}^{t''} (\boldsymbol{F}_{12}+\boldsymbol{F}_{21})\mathrm{d}t = 0. \tag{1.106}$$

As $\boldsymbol{F}_{12}$ ($\boldsymbol{F}_{21}$) is the only force acting upon the first (second) particle, we can use (1.103) to write

$$\boldsymbol{F}_{12} = \frac{\mathrm{d}m_1\boldsymbol{v}_1}{\mathrm{d}t}, \qquad \boldsymbol{F}_{21} = \frac{\mathrm{d}m_2\boldsymbol{v}_2}{\mathrm{d}t}, \tag{1.107}$$

and we get from (1.106) and (1.107)

$$[m_1\boldsymbol{v}_1+m_2\boldsymbol{v}_2]_{t'}^{t''} = 0, \tag{1.108}$$

or

$$\boldsymbol{p}_1'+\boldsymbol{p}_2' = \boldsymbol{p}_1''+\boldsymbol{p}_2'', \tag{1.109}$$

where the linear momentum $\boldsymbol{p}$ is defined by the equation

$$\boldsymbol{p} = m\boldsymbol{v}, \tag{1.110}$$

and where the primes (double primes) indicate the values of the quantities at t' (t'').

Equation (1.109) expresses the law of *conservation of (linear) momentum* which we have now shewn to hold for an isolated system of two interacting particles.

Let us now consider the motion of one particle under the influence of a force $\boldsymbol{F}$ and let us evaluate the integral

$$I = \int_{t'}^{t''} (\boldsymbol{F}\cdot \mathrm{d}\boldsymbol{x}). \tag{1.111}$$

Using (1.103), and writing $\mathrm{d}\boldsymbol{x} = \boldsymbol{v}\mathrm{d}t$, we have

$$I = \int_{t'}^{t''} (m\ddot{\boldsymbol{x}}\cdot\dot{\boldsymbol{x}})\mathrm{d}t = [\tfrac{1}{2}m\dot{\boldsymbol{x}}^2]_{t'}^{t''} = T''-T', \tag{1.112}$$

where we have introduced the *kinetic energy* T by the equation

$$T = \tfrac{1}{2}mv^2 = \tfrac{1}{2}m(\dot{\boldsymbol{x}}\cdot\dot{\boldsymbol{x}}), \tag{1.113}$$

and where a dot (two dots) denote here and henceforth differentiating once (twice) with respect to the time.

As $(\boldsymbol{F} \cdot \boldsymbol{v})$ is the work done per unit time by the force on the particle, we see that (1.112) expresses the fact that the total work done by the forces acting upon a particle during a time interval (t', t'') is equal to the change in the kinetic energy of the particle. From (1.112) one can prove the *conservation of energy*, provided the field of force is a *conservative* one. A field of force is conservative, if the forces can be derived from a potential function U by the equation

$$\boldsymbol{F} = -\nabla U, \tag{1.114}$$

where ∇ is the gradient operator with components $\partial/\partial x$, $\partial/\partial y$, and $\partial/\partial z$. In that case we have

$$\int_{t'}^{t''} (\boldsymbol{F} \cdot \mathrm{d}\boldsymbol{x}) = -\int_{t'}^{t''} (\nabla U \cdot \mathrm{d}\boldsymbol{x}) = -U'' + U', \tag{1.115}$$

or, using (1.111) and (1.112),

$$T' + U' = T'' + U''. \tag{1.116}$$

The potential U is called the *potential energy* and we see from (1.116) that, provided U does not depend explicitly on t, the *total energy* E, that is, the sum of the kinetic and the potential energy,

$$E = T + U, \tag{1.117}$$

is a *constant of motion*, that is, does not change in time during the motion of the particle.

We note from (1.115) that in the case of a conservative field of force, the integral on the left hand side does not depend on the path of the particle, but only on its position at the beginning and at the end of the time interval considered. If this were not the case, we could not, of course, have introduced a potential function. Indeed, one can define a conservative field of force by the requirement that the integral I of (1.111) depends only upon the positions of the particle at t' and t'', but not on the path traversed between t' and t''.

If we are dealing with a one-dimensional conservative system, the equations of motion can always be solved by quadrature. Since the energy is a constant of motion, we can write

$$E = \tfrac{1}{2}m\dot{x}^2 + U(x), \tag{1.118}$$

or,

$$\dot{x} = [(2/m)(E - U)]^{\frac{1}{2}} \tag{1.119}$$

from which we get

$$t-t_0 = \int_{x_0}^{x} \{2[E-U(x)]/m\}^{-\frac{1}{2}} \mathrm{d}x, \tag{1.120}$$

where x_0 is the position of the particle at t_0.

A simple example of such a case is that of a linear harmonic oscillator, which is defined by a potential energy U given by the equation

$$U = \tfrac{1}{2}ax^2. \tag{1.121}$$

The solution (1.120) is now of the form

$$t-t_0 = (m/a)^{\frac{1}{2}} \arcsin\,[x(a/2E)^{\frac{1}{2}}],$$

or

$$x = (2E/a)^{\frac{1}{2}} \sin 2\pi\nu(t-t_0), \qquad 2\pi\nu = (a/m)^{\frac{1}{2}}, \tag{1.122}$$

where ν is the frequency of the harmonic oscillator, and where for the sake of simplicity we have put x_0 equal to zero.

We note in (1.122) that the energy is proportional to the square of the amplitude of the oscillation.

2. CENTRAL FIELDS OF FORCE

In many physical systems the forces are of a special kind, namely, central forces. These are forces which are acting along the line connecting the body on which the force is acting with the body producing the field of force. If we restrict our discussion to the case of a single particle in an external field of force, a central field of force is one in which the force acting on the particle will be directed along the line connecting the particle and a fixed point, the centre of the force field. If we choose the origin at the centre of the field, the force $\boldsymbol{F}$ acting on the particle will be of the form

$$\boldsymbol{F} = f(x, y, z)\boldsymbol{x}. \tag{1.201}$$

In general such a field of force need not be conservative. If it is conservative, the function $f(x, y, z)$ occurring in (1.201) can depend on the distance r from the origin only,

$$\boldsymbol{F} = f(r)\boldsymbol{x}. \tag{1.202}$$

This can be shewn by using (1.114) for a conservative force and evaluating the components of $\boldsymbol{F}$. We get

$$F_x = -\frac{\partial U}{\partial x}, \qquad F_y = -\frac{\partial U}{\partial y}, \qquad F_z = -\frac{\partial U}{\partial z}, \tag{1.203}$$

and as (1.201) must hold, we have in spherical polars r, θ, φ,

$$-\frac{\partial U}{\partial x} = fr \sin\theta \cos\varphi, \quad -\frac{\partial U}{\partial y} = fr \sin\theta \sin\varphi, \quad -\frac{\partial U}{\partial z} = fr \cos\theta, \tag{1.204}$$

from which it follows that

$$\frac{\partial U}{\partial r} = -fr, \qquad \frac{\partial U}{\partial \theta} = 0, \qquad \frac{\partial U}{\partial \varphi} = 0. \tag{1.205}$$

From the last two equations, it follows that U depends on r only, and from the first equation we tnen see that (1.202) must hold, while

$$U(\boldsymbol{x}) = U(r) = -\int^r rf(r)\mathrm{d}r. \tag{1.206}$$

Particular examples of central forces are the isotropic harmonic oscillator and the Coulomb or gravitational force field. In the first case the potential is given by the equation

$$U = \tfrac{1}{2}ar^2, \tag{1.207}$$

and in the second case by the equation

$$U = -\kappa/r. \tag{1.208}$$

If the constant κ in (1.208) is positive we are dealing with an attractive force, while a negative κ corresponds to a repulsive force. Equation (1.208) leads, of course, to an inverse square force, that is, a force proportional to the inverse square of the distance from the origin.

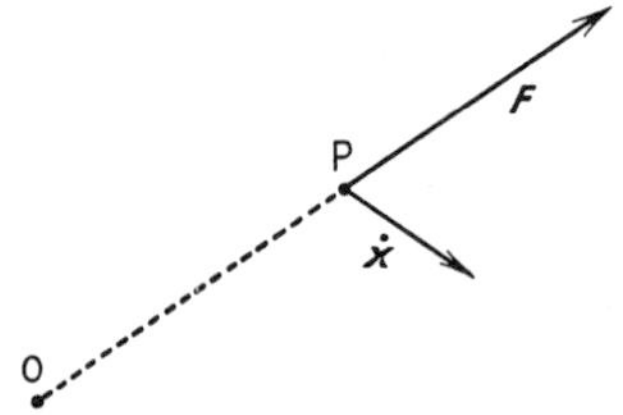

Fig. 1. Central field of force: O: centre of force field; P: position of particle; $\dot{\boldsymbol{x}}$: velocity vector of particle; $\boldsymbol{F}$: force acting upon particle.

The orbit of a particle on which the force (1.201) is acting will lie in a plane. This can be seen as follows (see fig. 1). According to Newton's second law, the acceleration of the particle will be in the direction of the force. Hence, both the acceleration and the velocity of the particle will be, and will remain, in the plane through the origin containing the velocity of the particle. Put differently, there will never be a component of the acceleration perpendicular to the plane through the origin and the velocity, so that the particle will stay in this plane, the *orbital plane*.

It is possible to prove that the orbit of a particle moving under the influence of a central field of force is a planar one by considering the *angular momentum* of the particle. The angular momentum with respect of the origin, $\boldsymbol{M}$, is defined by the equation

$$\boldsymbol{M} = [\boldsymbol{x} \wedge \boldsymbol{p}] = [\boldsymbol{x} \wedge m\dot{\boldsymbol{x}}]. \tag{1.209}$$

The angular momentum was in the past often called the moment of momentum, a very descriptive term. The nomenclature 'angular momentum' derives from the consideration of generalised coordinates which is given in the next chapter. The rate of change of $\boldsymbol{M}$ is given by the equation

$$\dot{\boldsymbol{M}} = [\dot{\boldsymbol{x}} \wedge m\dot{\boldsymbol{x}}]+[\boldsymbol{x} \wedge m\ddot{\boldsymbol{x}}] = 0+[\boldsymbol{x} \wedge f\boldsymbol{x}] = 0, \tag{1.210}$$

where we have used (1.105) and (1.201).

We have thus found for the case of a central force field a constant of motion, namely, $\boldsymbol{M}$, which is a vector, and thus really corresponds to three constants of motion. As $\boldsymbol{M}$ is constant, we see from (1.209) that the vector $\boldsymbol{x}$ will always lie in the fixed plane perpendicular to $\boldsymbol{M}$, which proves our statement that the particle orbit will be a planar one.

The problem of a particle in a central field of force can thus be reduced to a two-dimensional problem. The solutions of the original equations of motion, three second order differential equations, would contain six integration constants. Two of those can be chosen to be the direction cosines of $\boldsymbol{M}$, that is, they fix the normal to the orbital plane. In the remaining two-dimensional problem we are left with four integration constants which together with these two direction cosines make up the original six constants of motion. Of the remaining four, one will be the absolute magnitude of $\boldsymbol{M}$.

So far our conclusions are general and hold for forces which are not necessarily conservative. If the force field is conservative — and we shall from now on assume that this is the case — one of the last three constants of motion is the total energy and the other two will appear when the equations of motion are solved by quadrature, as can be done for a conservative central field.

Let us choose the z-axis of our system of coordinates along $\boldsymbol{M}$ and let us introduce polar coordinates r and θ in the xy-plane, that is, in the orbital plane, as follows

$$x = r\cos\theta, \qquad y = r\sin\theta. \tag{1.211}$$

Expressed in terms of x and y the equations of motion are of the form

$$m\ddot{x} = -\frac{x}{r}\frac{\mathrm{d}U}{\mathrm{d}r}, \qquad m\ddot{y} = -\frac{y}{r}\frac{\mathrm{d}U}{\mathrm{d}r}, \tag{1.212}$$

where U is given by (1.206), and the force by (1.202).

If M be the absolute magnitude of the angular momentum and E the energy, we have from (1.209) and (1.117),

$$M = m(x\dot{y}-y\dot{x}), \qquad E = \tfrac{1}{2}m(\dot{x}^2+\dot{y}^2)+U(r). \tag{1.213}$$

Equations (1.213) can, of course, be derived directly from (1.212) by quadrature.

Introducing r and θ we get instead of (1.213) the equations

$$M = mr^2\dot{\theta}, \tag{1.214}$$

$$E = \tfrac{1}{2}m\dot{r}^2+\tfrac{1}{2}mr^2\dot{\theta}^2+U(r). \tag{1.215}$$

Equation (1.214) describes the so-called *law of areas*. We know that M is a constant. To see the physical meaning of the right hand side of (1.214)

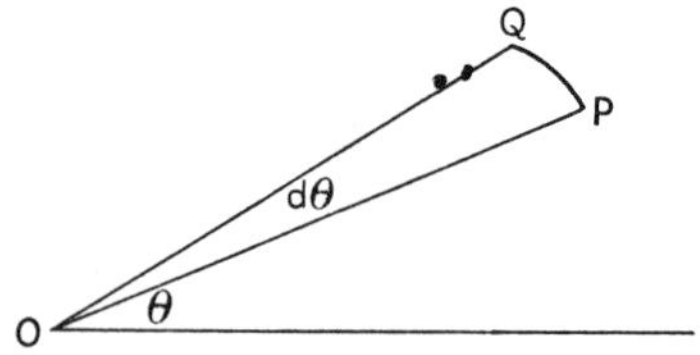

Fig. 2. The law of areas: O: centre of the force field; P: position of particle at time t; Q: position of particle at time $t+\mathrm{d}t$; θ: polar angle.

we consider fig. 2. Let the particle pass from P to Q during the time interval t, $t+\mathrm{d}t$. From the figure it is clear that during $\mathrm{d}t$ the radius vector sweeps through an area $\frac{1}{2}r^2\mathrm{d}\theta$. The areal velocity, defined as the area swept through per unit time, is thus equal to $\frac{1}{2}r^2\dot{\theta}$ and is constant according to (1.214). This is sometimes expressed as follows: The radius vector describes equal areas in equal times. For the case of the gravitational potential (1.208) this statement is known as Kepler's second law, but we have just seen that it is generally true for any central force, even for a non-conservative one.

Equation (1.215) gives an expression for the total energy in terms of polar coordinates, and we note that the kinetic energy T in planar polars is of the form

$$T = \tfrac{1}{2}m\dot{r}^2 + \tfrac{1}{2}mr^2\dot{\theta}^2, \tag{1.216}$$

where the two terms on the right hand side refer respectively to the radial and transverse motion of the particle.

Eliminating $\dot{\theta}$ from (1.214) and (1.215) we get

$$E = \tfrac{1}{2}m\dot{r}^2 + U(r) + \frac{M^2}{2mr^2}. \tag{1.217}$$

The last term on the right hand side of this equation might be called the centrifugal potential energy. The absolute magnitude $F_{\rm cf}$ of the force corresponding to this term would be given by the equation

$$F_{\rm cf} = -\frac{\mathrm{d}}{\mathrm{d}r}\frac{M^2}{2mr^2} = \frac{M^2}{mr^3} = \frac{mv_\perp^2}{r}, \tag{1.218}$$

where $v_\perp$ is the velocity component perpendicular to the radius vector ($M = mrv_\perp$). We see that we get for $F_{\rm cf}$ the usual expression for the centrifugal force acting on a particle which moves in a direction at an angle to the radius vector.

Formally (1.217) is of the same form as (1.118) for the one-dimensional case with an equivalent potential energy $U(r)+M^2/2mr^2$, that is, equal to the sum of the potential energy and centrifugal potential energy. We can thus solve this equation by quadrature, and the result is

$$t - t_0 = \int_{r_0}^{r} \frac{\mathrm{d}r}{\left[\dfrac{2E}{m} - \dfrac{2U(r)}{m} - \dfrac{M^2}{m^2r^2}\right]^{\frac{1}{2}}}, \tag{1.219}$$

where r_0 is the value of r at t_0. Combining (1.219) with (1.214) we get

$$\theta - \theta_0 = \int_{t_0}^{t} \frac{M\mathrm{d}t}{mr^2} = \int_{r_0}^{r} \frac{M\mathrm{d}r}{r^2\{2m[E-U(r)] - M^2r^{-2}\}^{\frac{1}{2}}}. \tag{1.220}$$

This equation gives us θ as a function of r, that is, the form of the orbit. The remaining two integration constants are now fixed: they are the values of θ and r at t_0.

Before considering quantitatively the orbit for a special choice of U, namely (1.208), we shall consider qualitatively different kinds of orbits which arise when $U(r)$ behaves as r^{-1} both for very large and for very small values of r, although not necessarily with the same coefficient. Such a potential is of interest in atomic problems. The last electron in an atom will feel a potential $-e/r$ at large distances from the nucleus — the nuclear charge being

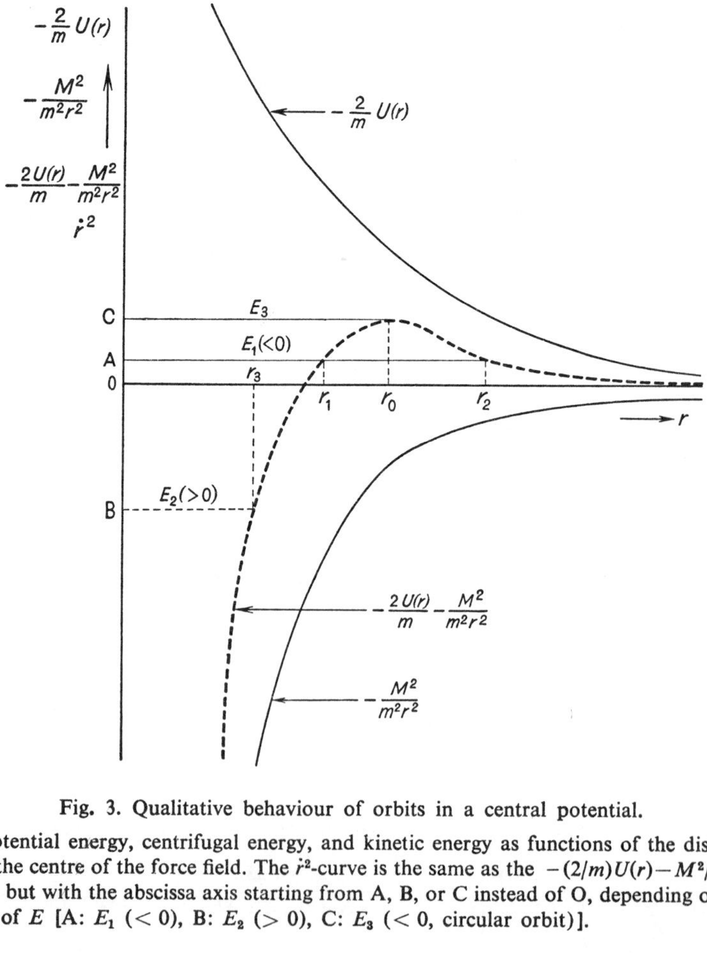

Fig. 3. Qualitative behaviour of orbits in a central potential.

a): Potential energy, centrifugal energy, and kinetic energy as functions of the distance from the centre of the force field. The $\dot{r}^2$-curve is the same as the $-(2/m)U(r)-M^2/m^2r^2$-curve, but with the abscissa axis starting from A, B, or C instead of O, depending on the value of E [A: E_1 (<0), B: E_2 (>0), C: E_3 (<0, circular orbit)].

screened by the other $Z-1$ electrons — and a potential $-Ze/r$ near the nucleus, when it is well inside the other electrons. In fig. 3 we have given first of all (Fig. 3a) $-2U(r)/m$, $-M^2/m^2r^2$, and their sum, as well as $\dot{r}^2$ as a function of r. This last function is obtained by rewriting (1.217) as follows

$$\dot{r}^2 = \frac{2E}{m} - \frac{2U(r)}{m} - \frac{M^2}{m^2r^2}; \tag{1.221}$$

it must be noted that in fig. 3a the abscissa axis is different for different

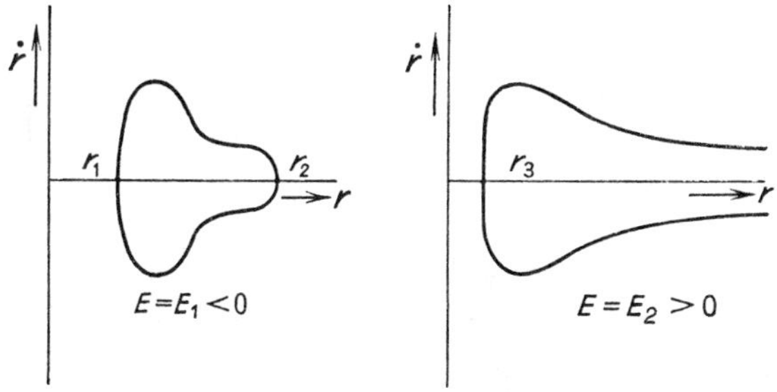

Fig. 3. b): Radial velocity as function of distance from centre of force field.

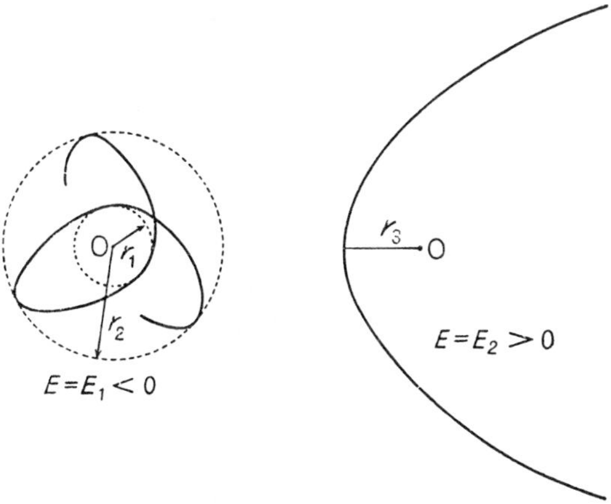

Fig. 3. c): Orbits for negative and positive energy values: O: centre of force field.

values of E in the $\dot{r}^2$ against r diagram. In fig. 3b we have plotted $\dot{r}$ as a function of r for two energy values E_1 and E_2 of which one is positive and the other one negative. It can be seen both from fig. 3a and from fig. 3b that for negative values of E the orbit will be one for which the particle does not disappear to infinity but is moving between two values of r, r_1 and r_2. It will thus in general be a rosette of the general shape shewn in fig. 3c. In the limiting case (E_3 in fig. 3a) where the r-axis just touches the $\dot{r}^2(r)$ curve we have a circular orbit. If E is positive, the orbit is an open one, and the particle has sufficient energy to disappear to infinity. This difference between open and closed orbits depending on whether E is positive or negative will crop up again presently, when we discuss the $1/r$-potential [see the discussion of (1.240)].

In many cases of interest the equations of motion simplify when we introduce $\sigma = r^{-1}$ instead of r as our coordinate. This substitution is the basis of Binet's method which is especially useful in the case where U is given by (1.208).

Let us first of all consider a decomposition of the force F acting on the particle into two components $F_{\|}$ and $F_{\perp}$ along and at right angles to the radius vector. We shall express this decomposition by using the Argand diagram of complex numbers,

$$m(\ddot{x}+\mathrm{i}\ddot{y}) = (F_{\|}+\mathrm{i}F_{\perp})\mathrm{e}^{\mathrm{i}\theta}. \tag{1.222}$$

If we write (1.211) in the composite form

$$x+\mathrm{i}y = r\mathrm{e}^{\mathrm{i}\theta}, \tag{1.223}$$

we find for the left hand side of equation (1.222)

$$m(\ddot{x}+\mathrm{i}\ddot{y}) = m(\ddot{r}-r\dot{\theta}^2+\mathrm{i}r\ddot{\theta}+2\mathrm{i}\dot{r}\dot{\theta})\mathrm{e}^{\mathrm{i}\theta}, \tag{1.224}$$

and hence

$$F_{\perp} = m(r\ddot{\theta}+2\dot{r}\dot{\theta}) = \frac{m}{r}\frac{\mathrm{d}}{\mathrm{d}t}\, r^2\dot{\theta}, \tag{1.225}$$

$$F_{\|} = m(\ddot{r}-r\dot{\theta}^2). \tag{1.226}$$

As we are dealing with central forces, $F_{\perp}$ vanishes, and (1.225) leads to (1.214), while from equations (1.212) it follows that the left hand side of (1.224) is equal to $-(\mathrm{d}U/\mathrm{d}r)\mathrm{e}^{i\theta}$, so that we can rewrite (1.226) in the form

$$F_{\|} = -\frac{\mathrm{d}U}{\mathrm{d}r} = m(\ddot{r}-r\dot{\theta}^2). \tag{1.227}$$

We may add that it follows conversely from (1.225) that if $r^2\dot{\theta}$ is constant, $F_\perp$ will vanish and we are dealing with a central force.

Let us now evaluate $\mathrm{d}^2\sigma/\mathrm{d}\theta^2$. Using (1.214) we find

$$\frac{\mathrm{d}\sigma}{\mathrm{d}\theta} = \frac{\mathrm{d}}{\mathrm{d}\theta}\frac{1}{r} = \frac{\mathrm{d}}{\mathrm{d}t}\frac{1}{r}\Big/\frac{\mathrm{d}\theta}{\mathrm{d}t} = -\frac{\dot{r}}{r^2\dot{\theta}} = -\frac{m\dot{r}}{M}, \tag{1.228}$$

and hence, using (1.227),

$$\frac{\mathrm{d}^2\sigma}{\mathrm{d}\theta^2} = \frac{\mathrm{d}}{\mathrm{d}t}\left(-\frac{m\dot{r}}{M}\right)\Big/\frac{\mathrm{d}\theta}{\mathrm{d}t} = -\frac{m\ddot{r}}{M\dot{\theta}} = \frac{1}{M\dot{\theta}}\left[\frac{\mathrm{d}U}{\mathrm{d}r} - mr\dot{\theta}^2\right]$$

$$= -\sigma + \frac{1}{M\dot{\theta}}\frac{\mathrm{d}U}{\mathrm{d}r} = -\sigma + \frac{\mathrm{d}U}{\mathrm{d}\sigma}\frac{\mathrm{d}\sigma}{\mathrm{d}r}\frac{m}{M^2\sigma^2} = -\sigma - \frac{m}{M^2}\frac{\mathrm{d}U}{\mathrm{d}\sigma},$$

or,

$$\frac{\mathrm{d}^2\sigma}{\mathrm{d}\theta^2} + \sigma = -\frac{m}{M^2}\frac{\mathrm{d}U}{\mathrm{d}\sigma}. \tag{1.229}$$

So far we have not yet made any special choice for $U(r)$. We shall now, however, consider especially the case where U is given by (1.208). In that case $U(\sigma) = -\kappa\sigma$, and (1.229), which is the equation for the orbit, becomes especially simple,

$$\frac{\mathrm{d}^2\sigma}{\mathrm{d}\theta^2} + \sigma = \frac{m\kappa}{M^2},$$

with the solution

$$\sigma = \frac{1}{r} = \frac{m\kappa}{M^2} + A\cos(\theta - \theta_0). \tag{1.230}$$

This orbit is a conic section (see fig. 4). In fig. 4 the length of OS is A^{-1}. It is easily seen that

$$|\mathrm{OQ}| = r\cos(\theta-\theta_0), \qquad |\mathrm{PR}| = [1 - Ar\cos(\theta-\theta_0)]/A,$$

so that, using (1.230), we have

$$\frac{|\mathrm{OP}|}{|\mathrm{PR}|} = \frac{rA}{1 - Ar\cos(\theta-\theta_0)} = \frac{AM^2}{m\kappa} = \text{constant}. \tag{1.231}$$

The locus of all points P satisfying (1.231) is a conic section, proving our statement. We shall discuss in a moment under what circumstances the orbit is an ellipse, a parabola, or a hyperbola. Equation (1.230) could have been derived directly from (1.220) after substituting expression (1.208) for $U(r)$.

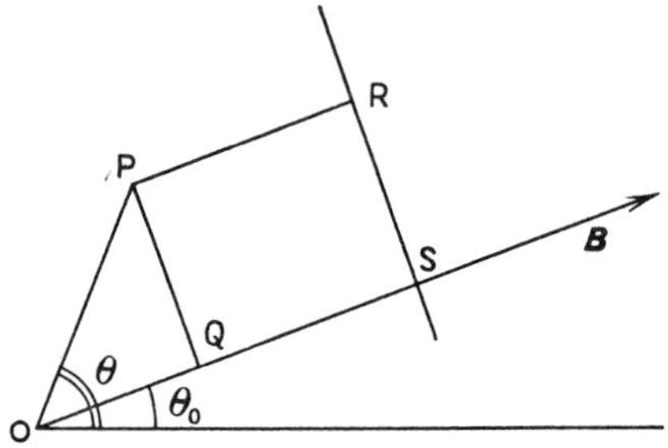

Fig. 4. The locus of P is a conic section, if |OP|/|PR| is constant. O: centre of force field; P: position of particle; the angles PQO, PRS, and RSQ are all right angles; the vector $\boldsymbol{B}$ is the integration constant of equation (1.236).

We wish to use a different method to derive (1.230). This method is based upon vector calculus and is specially suited to the $1/r$ potential. The equations of motion (1.105) have the form

$$m\ddot{\boldsymbol{x}} = -\frac{\kappa \boldsymbol{x}}{r^3}, \tag{1.232}$$

where we have used (1.114) and (1.208). We use the fact that the angular momentum vector $\boldsymbol{M}$ given by (1.209) is a constant vector, and we introduce a vector $\boldsymbol{Q}$ by the equation

$$\boldsymbol{Q} = [\boldsymbol{M} \wedge \dot{\boldsymbol{x}}]. \tag{1.233}$$

As $\boldsymbol{Q}$ is perpendicular to $\boldsymbol{M}$, it will lie in the orbital plane. If we use (1.209), the fact that $\boldsymbol{M}$ is constant, equation (1.232), and the relation $(\boldsymbol{x} \cdot \dot{\boldsymbol{x}}) = r\dot{r}$, which follows from $r^2 = (\boldsymbol{x} \cdot \boldsymbol{x})$, we find for the rate of change of $\boldsymbol{Q}$

$$\dot{\boldsymbol{Q}} = [\boldsymbol{M} \wedge \ddot{\boldsymbol{x}}] = -\frac{\kappa}{r^3}[[\boldsymbol{x} \wedge \dot{\boldsymbol{x}}] \wedge \boldsymbol{x}] = -\kappa\left[\frac{\dot{\boldsymbol{x}}}{r} - \boldsymbol{x}\frac{\dot{r}}{r^2}\right] = -\kappa\frac{\mathrm{d}}{\mathrm{d}t}\frac{\boldsymbol{x}}{r},$$

or

$$\dot{\boldsymbol{Q}} = -\kappa\dot{\boldsymbol{x}}_0, \tag{1.234}$$

where $\boldsymbol{x}_0$ is the unit vector in the direction of the radius vector,

$$\boldsymbol{x}_0 = \boldsymbol{x}/r. \tag{1.235}$$

Equation (1.233) can be integrated and leads to the expression

$$\boldsymbol{Q}+\kappa \boldsymbol{x}_0 = -\boldsymbol{B}, \tag{1.236}$$

where $\boldsymbol{B}$ is a constant vector.

Taking the scalar product of equation (1.236) and $\boldsymbol{x}_0$ we get

$$-\frac{M^2}{mr}+\kappa = -B\cos(\theta-\theta_0) \tag{1.237}$$

where $B = |\boldsymbol{B}|$, where we have assumed that the vector $\boldsymbol{B}$ makes an angle θ_0 with the x-axis in the orbital plane (see fig. 4) and the vector $\boldsymbol{x}_0$ an angle θ [compare (1.211) and fig. 2] and where we have used the equation

$$(\boldsymbol{Q}\cdot\boldsymbol{x}_0) = \left(\frac{\boldsymbol{x}}{r}\cdot[\boldsymbol{M}\wedge\dot{\boldsymbol{x}}]\right) = \frac{1}{r}(\boldsymbol{M}\cdot[\dot{\boldsymbol{x}}\wedge\boldsymbol{x}]) = -\frac{M^2}{mr}. \tag{1.238}$$

Equation (1.237) is the same as (1.230), if we put $B = M^2A/m$.

We shall now take the square of the two sides of equation (1.236). As $\boldsymbol{M}$ is perpendicular to $\dot{\boldsymbol{x}}$, $\boldsymbol{Q}^2$ will be equal to $M^2\dot{\boldsymbol{x}}^2$, and hence we have

$$M^2\dot{\boldsymbol{x}}^2+\kappa^2-\frac{2\kappa M^2}{mr} = B^2, \tag{1.239}$$

where we have used (1.238).

Equation (1.239) can be written in the following form

$$E = \tfrac{1}{2}m\dot{\boldsymbol{x}}^2-\frac{\kappa}{r} = \frac{(B^2-\kappa^2)m}{2M^2}, \tag{1.240}$$

and we have obtained the energy equation. We see that in the case of positive κ the sign of E, and hence the character of the orbit is governed by the magnitude of B. (If κ is negative, E is always positive and the orbit always hyperbolic.) If $B > \kappa$, the particle can disappear to infinity with a finite velocity, and we have a hyperbola (curve 1 in fig. 5). We must emphasise here that we get, of course, only one of the two branches of the hyperbola as r is essentially non-negative. The origin is the interior focus, if $\kappa > 0$, and the exterior focus, if $\kappa < 0$. If $B = \kappa$, the velocity at infinity will be zero, and the orbit a parabola (curve 2), while for $B < \kappa$ the orbit is an ellipse (curve 3). If $B = 0$, the energy attains its smallest value compatible with a given value of M, and the orbit is circular.

Using the relation $M^2A/m = B$ which reduces (1.237) to (1.230) we see that the conditions $B > \kappa$, $B = \kappa$, and $B < \kappa$ correspond to ratios |OP| to |PR| in (1.231) larger than, equal to, and less than unity, also shewing that these conditions refer to hyperbolic, parabolic, and elliptic orbits.

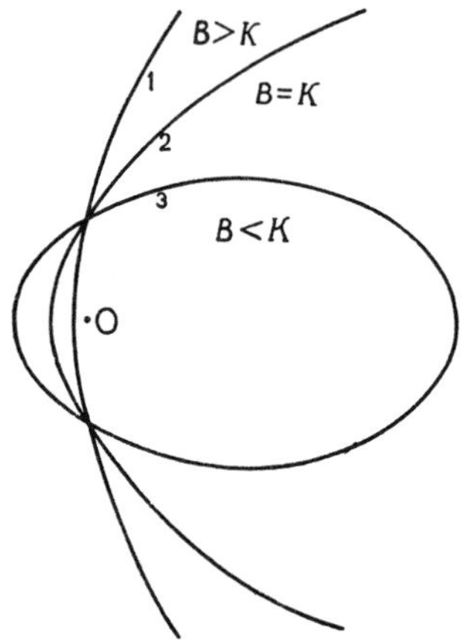

Fig. 5. Possible orbits of a particle of given angular momentum: O: centre of field of force.

From (1.237) we see that the value of r for $\theta = \theta_0 \pm \frac{1}{2}\pi$, often called the *parameter*, p, of the conical section, is independent of B, that is, of the energy, and depends only on the angular momentum,

$$p = \frac{M^2}{m\kappa}, \tag{1.241}$$

a fact which played an important role in the discussion of the old quantum theory.

Let us finally consider the apocentre and pericentre values of r (r_{max} and r_{min}) in the case of an ellipse ($E < 0$). They occur for $\theta = \theta_0 + \pi$ and $\theta = \theta_0$ respectively, and we have

$$r_{min} = \frac{M^2}{m(\kappa + B)}, \qquad r_{max} = \frac{M^2}{m(\kappa - B)}, \tag{1.242}$$

and the semi-major axis a will satisfy the equation

$$2a = r_{min} + r_{max} = -\frac{\kappa}{E}, \quad \text{or,} \quad E = -\frac{\kappa}{2a}. \tag{1.243}$$

We finally want to express the eccentricity ε of the ellipse and the period τ of the particle in its orbit in terms of E, κ, and M. From (1.242) and (1.240) we get for ε the equation

$$\varepsilon = \frac{r_{\max}-r_{\min}}{r_{\max}+r_{\min}} = \frac{B}{\kappa} = \left[\frac{2M^2E}{m\kappa^2}+1\right]^{\frac{1}{2}}. \tag{1.244}$$

We see again that $B/\kappa < 1$ corresponds to $\varepsilon < 1$, that is, to an ellipse; that $B = \kappa$ corresponds to $\varepsilon = 1$, that is, to a parabola; and that $B/\kappa > 1$ corresponds to $\varepsilon > 1$, that is, to a hyperbola. Using the relation $\varepsilon = B/\kappa$ and (1.241) we can write (1.237) in the familiar form

$$\frac{p}{r} = 1+\varepsilon \cos(\theta-\theta_0). \tag{1.245}$$

We saw earlier that Kepler's second law is on the one hand equivalent to stating that the angular momentum M is constant, and on the other hand is the same as the law of areas. In fact, we saw from (1.214) that $M/2m$ is the areal velocity. The total area of the ellipse which is swept through during a single period τ is equal to πab, that is, to $\pi a^2(1-\varepsilon^2)^{\frac{1}{2}}$, where b is the semi-minor axis, so that we have

$$\pi a^2(1-\varepsilon^2)^{\frac{1}{2}}/\tau = M/2m,$$

and, using (1.243) and (1.244)

$$\frac{\tau^2}{a^3} = \frac{4\pi^2 m}{\kappa}, \tag{1.246}$$

which reduces to Kepler's third law if we put $\kappa = GmM_\infty$, where G is the gravitational constant and M_∞ the mass of the (infinitely heavy) central mass. If the central mass is finite, equation (1.246) must be corrected as we shall see presently.

At the beginning of this section we defined a central force as one where the force exerted by one particle upon another one would be directed along the line connecting the two particles. This, of course, is the physical situation which one will often encounter, and we shall shew now that this situation will lead to the problem of a single particle in an external central field of force, that is, to the problem which we have just discussed at some length.

Let m_1 and m_2 be the masses of the two particles and let the force exerted by the particles upon each other be derivable from a potential $U(r_{12})$, where

r_{12} is the distance apart of the two particles. One can easily shew by methods similar to the one which led to (1.206) that if the forces considered are conservative they can, indeed, be derived from a potential which is a function of r_{12} only.

The equations of motion for the two particles are

$$m_1\ddot{\boldsymbol{x}}_1 = -\nabla_1 U, \qquad m_2\ddot{\boldsymbol{x}}_2 = -\nabla_2 U, \tag{1.247}$$

and introducing centre of mass and relative coordinates through the relations

$$\boldsymbol{X} = \frac{m_1\boldsymbol{x}_1+m_2\boldsymbol{x}_2}{m_1+m_2}, \tag{1.248a}$$

$$\boldsymbol{x} = \boldsymbol{x}_1-\boldsymbol{x}_2, \tag{1.248b}$$

we find from (1.247)

$$(m_1+m_2)\ddot{\boldsymbol{X}} = 0, \tag{1.249}$$

$$\mu\ddot{\boldsymbol{x}} = -\nabla U, \tag{1.250}$$

where μ is the *reduced mass,*

$$\mu = \frac{m_1 m_2}{m_1+m_2} \tag{1.251}$$

Equations (1.249) and (1.250) express the fact that the motion of the two particles can be described as a superposition of the centre of mass motion which in this case is a free particle motion (1.249) and the relative motion (1.250) which is the motion of a particle with the reduced mass moving in the central field of force determined by the given potential energy. If one mass is much larger than the other, the reduced mass is approximately equal to the lighter of the two masses. This accounts for the fact that Kepler's third law holds to such a good approximation. More exactly it should be

$$\frac{\tau^2}{a^3} = \frac{4\pi^2\mu}{\kappa} = \frac{4\pi^2}{G(m+M_c)},$$

where M_c is the mass of the central body (the sun in the planetary system). If $M_c \gg m$, the right hand side reduces to a constant, $4\pi^2/GM_c$.

Let us end this section by considering scattering processes in a central field of force. The fact that this field will often be produced by another particle only means that we must use the reduced mass instead of the free particle mass in our considerations. In scattering one is not so much concerned with the actual processes which occur when the scattered particle is near the scattering particle as with the final outcome of the scattering process. Put differently, one is interested in such quantities as the scattering cross section or the probability that scattering over a given angle will take place. The initial conditions are determined by the energy and angular momentum of the incident particle. Let v be its speed at infinity, and let the impact parameter, that is, the nearest distance at which the particle would pass the scattering

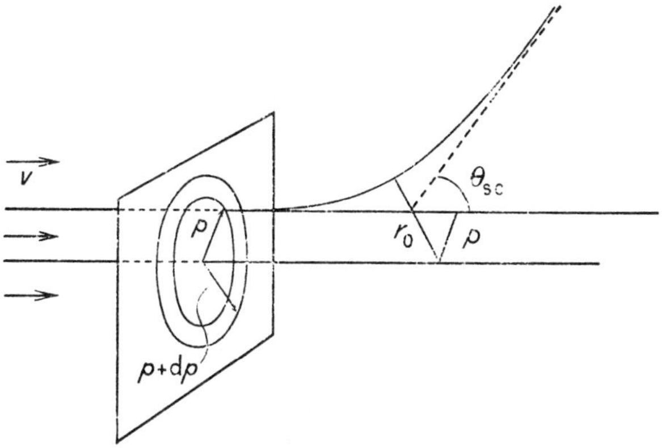

Fig. 6. Scattering by a repulsive central force: r_0: distance of nearest approach; p: impact parameter; θ_{sc}: scattering angle.

centre, if it were not deflected, be p (see fig. 6). In terms of v and p, we have

$$E = \tfrac{1}{2}mv^2, \qquad M = mvp, \tag{1.252}$$

and we get for the orbit from (1.220) the expression

$$\theta - \theta_0 = \int_{r_0}^{r} \frac{pv\,dr}{r^2[v^2-(2U/m)-(p^2v^2/r^2)]^{\frac{1}{2}}}\,. \tag{1.253}$$

If we take for r_0 the distance of nearest approach, that is, the value of r for which $\dot{r} = 0$, or for which

$$v^2 - \frac{2U(r)}{m} - \frac{p^2v^2}{r^2} = 0, \tag{1.254}$$

we get for the scattering angle θ_{sc} from (1.253) (see fig. 6, and note that θ

changes from π to θ_{sc} when r goes from ∞ to r_0 and again to ∞),

$$\pi-\theta_{sc} = 2\int_{r_0}^{\infty} \frac{pv\,dr}{r^2[v^2-(2U/m)-(p^2v^2/r^2)]^{\frac{1}{2}}}. \tag{1.255}$$

The differential cross section $d\sigma$ for scattering over an angle between θ_{sc} and $\theta_{sc}+d\theta_{sc}$ is defined as the ratio of the number of particles scattered over such angles to the incident intensity which is the total number of particles incident on the scattering centre per unit area, assuming that the incident particles form a uniform beam. From (1.255) we see that the scattering angle θ_{sc} is a function of the impact parameter p, and instead of asking for the number of particles which are scattered over a given angle θ_{sc}, we can ask for the number of particles which will be incident with an impact parameter between p and $p+dp$. This number is clearly equal to $2\pi Ip\,dp$ (I: incident intensity) as can be seen from fig. 6. We get thus for the differential cross section

$$d\sigma = 2\pi Ip\,dp, \tag{1.256}$$

and by solving (1.255) we can find the differential cross section in terms of θ_{sc}. The total scattering cross section is obtained by integrating this expression over all possible values of θ_{sc} from 0 to π.

Let us now consider especially the case where U is given by the expression $+\kappa/r$, and let us introduce the notation

$$u = \frac{p}{r}, \qquad b = \frac{2\kappa}{mv^2} = \gamma p. \tag{1.257}$$

The variable u is essentially Binet's variable, while b is the distance at which the absolute magnitude of the potential energy is equal to the kinetic energy at infinity; this means that b is the distance of nearest approach of a particle with speed v and zero angular momentum to the repulsive scattering centre. Using these variables we get from (1.255) ($u_0 = p/r_0$)

$$\tfrac{1}{2}\pi-\tfrac{1}{2}\theta_{sc} = \int_0^{u_0} \frac{du}{[1-\gamma u-u^2]^{\frac{1}{2}}} = \tfrac{1}{2}\pi-\arcsin\frac{\gamma}{(\gamma^2+4)^{\frac{1}{2}}},$$

or

$$\tan\tfrac{1}{2}\theta_{sc} = \tfrac{1}{2}\gamma. \tag{1.258}$$

One usually expresses $d\sigma$ as the product of the element of solid angle

$d\Omega \equiv 2\pi \sin\theta_{sc} d\theta_{sc}$ corresponding to scattering angles between θ_{sc} and $\theta_{sc} + d\theta_{sc}$, and the scattering cross section $R(\theta_{sc})$,

$$d\sigma = R(\theta_{sc}) d\Omega = R(\theta_{sc}) \cdot 2\pi \sin\theta_{sc} d\theta_{sc}. \tag{1.259}$$

For the scattering cross section we find from (1.256), (1.258) and (1.259)

$$R(\theta_{sc}) = \tfrac{1}{4} b^2 \operatorname{cosec}^4 \tfrac{1}{2}\theta_{sc}, \tag{1.260}$$

the famous Rutherford scattering formula. For this particular scattering process, the total cross section σ_{tot} ($= \int d\sigma = \int R d\Omega$) diverges. This is due to the fact that the Coulomb potential has such a 'long tail'. This is another way of saying that the Coulomb potential decreases so slowly with increasing distance from the force centre that θ_{sc} remains appreciable, even when p becomes very large; another way of stating this is to say that the range of the Coulomb forces is infinite.

In considering scattering processes we have so far assumed the scattering centre to be fixed. In actual scattering experiments one is concerned with the scattering of one particle by another particle. In that case we have a situation like the one discussed earlier in this section, namely, a case of two particles which are exerting forces upon one another. We saw then that the relative motion is the same as if the centre of mass were at rest and a particle with the reduced mass were moving in the field of force produced by the potential from which the interparticle forces can be derived. The substitution of the reduced mass for the mass of the scattered particle is a trivial operation. The fact that the formulae which we have just derived for scattering cross sections apply to the case where the scattering centre is at rest while in actual scattering experiments the target particle is at rest is, however, of great importance for the analysis of scattering data. As the target particle will be at rest, the centre of mass, that is, the scattering centre, will be moving and the scattering data which are obtained in the laboratory frame of reference must first be converted to data in the centre of mass frame of reference before our formulae can be applied.

3. SYSTEMS OF PARTICLES

We shall now consider systems consisting of several mass points. We shall assume that the force $\boldsymbol{F}_i$ on the ith particle can be derived from a potential U, as follows,

$$\boldsymbol{F}_i = -\nabla_i U, \tag{1.301}$$

so that the equations of motion are

$$m_i \ddot{\boldsymbol{x}}_i = \boldsymbol{F}_i = -\nabla_i U. \qquad (1.302)$$

The first consequence of (1.301) is that our system is conservative, that is, that the total energy E given by the equation

$$E = T + U = \tfrac{1}{2} \sum_i m_i \dot{\boldsymbol{x}}_i^2 + U \qquad (1.303)$$

is a constant, since

$$\frac{\mathrm{d}}{\mathrm{d}t}(T+U) = \sum_i m_i (\ddot{\boldsymbol{x}}_i \cdot \dot{\boldsymbol{x}}_i) + \sum_i (\dot{\boldsymbol{x}}_i \cdot \nabla_i U) = 0.$$

We shall further specify our system by assuming that the total potential energy is the sum of the potential energies of pairs of particles, and that these potential energies are functions of the distance apart of the two particles, r_{ij}, only,

$$U = \tfrac{1}{2} \sum_{i,j} U_{ij}(r_{ij}), \qquad U_{ij} = U_{ji}, \qquad U_{ii} = 0. \qquad (1.304)$$

One of the consequences of this equation is that the force on each particle is the vector sum of the (central) forces exerted on that particle by all the other particles; this means that the forces are *additive*. To prove this we see first of all that the force $\boldsymbol{F}_{ij}$ exerted by particle j on particle i is given by the equation

$$\boldsymbol{F}_{ij} = -\nabla_i U_{ij} = -\frac{\mathrm{d}U_{ij}}{\mathrm{d}r_{ij}} \nabla_i r_{ij} = -\frac{\mathrm{d}U_{ij}}{\mathrm{d}r_{ij}} \frac{\boldsymbol{x}_{ij}}{r_{ij}}, \qquad \boldsymbol{x}_{ij} = \boldsymbol{x}_i - \boldsymbol{x}_j. \qquad (1.305)$$

The total force on the i-th particle is given by the equation

$$\boldsymbol{F}_i = -\nabla_i U = -\sum_j \nabla_i U_{ij} = \sum_j \boldsymbol{F}_{ij}, \qquad (1.306)$$

which proves our statement.

If the potential energy is of the form (1.304) so that the only forces acting on the particles in the system are the inter-particle forces, not only the total energy, but also the total linear momentum $\boldsymbol{P}$ and the total angular momentum $\boldsymbol{M}$ of the system are constants of motion. We can prove that as follows. The total linear momentum and its rate of change are given by the equations

$$\boldsymbol{P} = \sum_i m_i \dot{\boldsymbol{x}}_i, \qquad (1.307)$$

$$\frac{\mathrm{d}\boldsymbol{P}}{\mathrm{d}t} = \sum_i m_i \ddot{\boldsymbol{x}}_i = -\sum_{i,j} \frac{\mathrm{d}U_{ij}}{\mathrm{d}r_{ij}} \frac{\boldsymbol{x}_{ij}}{r_{ij}}, \qquad (1.308)$$

where we have used (1.305), (1.306) and (1.302). In (1.308) i and j are dummy indices and if we interchange them, that is, call $i : j$ and vice versa, the last double sum should not change. On the other hand, if we interchange in that double sum i and j the sum changes its sign because $\boldsymbol{x}_{ij} = -\boldsymbol{x}_{ji}$. As the sum is thus equal to its own negative it must vanish which proves that $\boldsymbol{P}$ is constant.

Similarly we have for $\boldsymbol{M}$ and its rate of change

$$\boldsymbol{M} = \sum_i m_i[\boldsymbol{x}_i \wedge \dot{\boldsymbol{x}}_i], \tag{1.309}$$

$$\frac{\mathrm{d}\boldsymbol{M}}{\mathrm{d}t} = \sum_i m_i[\boldsymbol{x}_i \wedge \ddot{\boldsymbol{x}}_i] = -\sum_{i,j} \frac{\mathrm{d}U_{ij}}{\mathrm{d}r_{ij}}\left[\boldsymbol{x}_i \wedge \frac{\boldsymbol{x}_{ij}}{r_{ij}}\right] = \sum_{i,j} \frac{\mathrm{d}U_{ij}}{\mathrm{d}r_{ij}} \frac{[\boldsymbol{x}_i \wedge \boldsymbol{x}_j]}{r_{ij}}, \tag{1.310}$$

where we have used (1.305), (1.306), (1.302), and the relation $\boldsymbol{x}_{ij} = \boldsymbol{x}_i - \boldsymbol{x}_j$. Using the same arguments as those which proved that $\boldsymbol{P}$ is constant we prove that $\boldsymbol{M}$ is constant.

It is interesting to prove directly that $\boldsymbol{P}$ and $\boldsymbol{M}$ are constant by using Newton's third law ($\boldsymbol{F}_{ij} = -\boldsymbol{F}_{ji}$), Newton's second law, and a geometric argument. We leave this as an exercise to the reader.

An important theorem — both in classical and in quantum mechanics — is the *virial theorem* which deals with the time average of Clausius' virial $\mathscr{V}$, defined by the equation

$$\mathscr{V} = \sum_i (\boldsymbol{F}_i \cdot \boldsymbol{x}_i). \tag{1.311}$$

The virial can be written as follows

$$\mathscr{V} = \sum_i m_i(\ddot{\boldsymbol{x}}_i \cdot \boldsymbol{x}_i) = \frac{\mathrm{d}^2}{\mathrm{d}t^2}\left(\sum_i \tfrac{1}{2}m_i r_i^2\right) - 2T. \tag{1.312}$$

If we are dealing with a system which remains together, that is, for which none of the $\boldsymbol{x}_i$ or the $\dot{\boldsymbol{x}}_i$ ever becomes infinitely large, the time average of the first term on the right hand side of (1.312) will be zero, and we have in that case

$$\overline{2T + \mathscr{V}} = 0, \tag{1.313}$$

where the bar indicates a time average

$$\bar{f} = \lim_{T \to \infty} T^{-1} \int_t^{t+T} f(t)\,\mathrm{d}t.$$

If (1.301) holds, we can also write

$$\mathscr{V} = -\sum_i (\mathbf{x}_i \cdot \nabla_i U). \tag{1.314}$$

In the often realised case where U is a homogeneous function of the coordinates of degree g, we can use Euler's theorem for such functions to write for the right hand side of (1.314) $-gU$. Combining (1.313), (1.314) and (1.303) we get

$$2\overline{T} - g\overline{U} = 0, \qquad \overline{T} + \overline{U} = E,$$

or

$$\overline{U} = 2E/(g+2), \qquad \overline{T} = gE/(g+2). \tag{1.315}$$

In the particular case where $g = -1$ which is the case corresponding, for instance, to the potential (1.208) we find, first of all, that as T is positive definite, E must be negative, if the system is to stay together. Put differently, the condition that the virial theorem (1.313) holds leads to a negative value of E in this case.

If we are dealing with a circular orbit of radius a, $\overline{U} = U = -\kappa/a$, and we find from (1.315) $E = -\kappa/2a$ in accordance with (1.243). In the case of an elliptic orbit (1.243) still holds, but now equations (1.315) lead to the statement that the average value of U is equal to $-\kappa/a$, while the average value of the kinetic energy is equal to $\kappa/2a$.

CHAPTER 2

THE LAGRANGIAN EQUATIONS OF MOTION

In this chapter we first of all discuss how constraints can be introduced as limiting cases of ordinary potential energies. D'Alembert's principle is then discussed and the Lagrangian equations of the first kind are derived and applied to a number of simple cases. Hamilton's variational principle is derived and used to obtain the Lagrangian equations of the second kind, after generalised coordinates have been introduced. After that cyclic coordinates, the Routh function and hidden masses are discussed. This is followed by a brief discussion of non-holonomic non-integrable constraints and of velocity dependent potentials, with special reference to motion in an electromagnetic field, and the chapter ends with a brief discussion of the relation between infinitesimal transformations and conservation laws.

1. CONSTRAINTS

Let us consider a system of N particles and let $\boldsymbol{x}_i$ be the position coordinate of the i-th particle ($i = 1, \ldots, N$). Let $\boldsymbol{F}_i$ be the force acting on the i-th particle. This force can be split into two parts, one, $\boldsymbol{F}_i^{\text{int}}$, due to the other particles in the system and one, $\boldsymbol{F}_i^{\text{ext}}$ due to external fields of force,

$$\boldsymbol{F}_i = \boldsymbol{F}_i^{\text{int}} + \boldsymbol{F}_i^{\text{ext}}. \tag{2.101}$$

The system considered has $3N$ degrees of freedom and provided we know the $\boldsymbol{F}_i^{\text{int}}$ and the $\boldsymbol{F}_i^{\text{ext}}$ we can, in principle, solve the relevant equations of motion. Often, however, it is impossible to find all the $\boldsymbol{F}_i^{\text{int}}$ and $\boldsymbol{F}_i^{\text{ext}}$. Moreover, some of them may be of such a nature that the actual number of degrees of freedom is much less than $3N$. To see how this may come about, let us consider the following three potential energies,

$$U_1(x_1, x_2) = \tfrac{1}{2}a(x_1 - x_2 - l)^2; \tag{2.102}$$

$$\begin{aligned} U_2(x) &= 0, & |x| &< \tfrac{1}{2}L, \\ U_2(x) &= (2x - L)/b^2L, & \tfrac{1}{2}L \leqq |x| &\leqq \tfrac{1}{2}L(1+b), \\ U_2(x) &= 1/b, & \tfrac{1}{2}L(1+b) &< |x|; \end{aligned} \tag{2.103}$$

$$U_3(x, y, z) = \tfrac{1}{2}c(\alpha x+\beta y+\gamma z-d)^2, \qquad \alpha^2+\beta^2+\gamma^2 = 1. \quad (2.104)$$

The first potential refers to a one-dimensional two-particle system, the second to a one-dimensional one-particle system, and the third one to a three-dimensional one-particle system. It will be assumed that for the systems considered these potentials are the only ones present.

For the first case we use the relative coordinate x_1-x_2 and the centre of mass coordinate $(m_1x_1+m_2x_2)/(m_1+m_2)$ as new variables. As we saw in § 1.2 the centre of mass coordinate will describe a uniform motion while for the relative coordinate we get [compare (1.122)]

$$x_1-x_2 = l+\sqrt{\frac{2E}{a}}\sin\sqrt{\frac{a}{\mu}}(t-t_0), \quad (2.105)$$

where E is the energy of the relative motion and μ the reduced mass given by (1.251).

Let us now consider the limit as a tends to infinity for a given value of E. We see that the frequency of the oscillations of x_1-x_2 around the value l increases, while their amplitude decreases. In the limit we find

$$x_1-x_2 = l: \quad (2.106)$$

the system behaves as if the two particles are joined together at a fixed distance apart, l. To put it in modern language: the degree of freedom corresponding to the relative motion is *frozen in*. This freezing in of degrees of freedom is, of course, of great importance for quantum mechanical considerations. In classical mechanics, however, it is sufficient to say that the system is subject to a *constraint*; in the system under consideration this constraint is described by (2.106).

The potential U_1 can be generalised to three dimensions as follows

$$U_1' = \tfrac{1}{2}a'[\sqrt{(x_1-x_2)^2+(y_1-y_2)^2+(z_1-z_2)^2}-l]^2, \quad (2.107)$$

and the constraint (2.106) is replaced by the constraint

$$|\boldsymbol{x}_1-\boldsymbol{x}_2| = l \quad (2.108)$$

when $a' \to \infty$. The proof of this constraint is straightforward and is left to the reader.

A different kind of constraint occurs for the system subject to the potential U_2 (see fig. 7a) which corresponds physically to two walls of height $1/b$.

The motion of the particle is one-dimensional and one can use (1.120) to find its position as a function of time. One finds that, if the energy E of the particle is less than $1/b$ it will move up and down between the two walls where it is reflected. When it reaches the point corresponding to A in fig. 7a, at time t_1, say, it will begin to decelerate until its velocity vanishes, at time t_2 say, (point B in fig. 7a); it then goes back, accelerating from B to A until

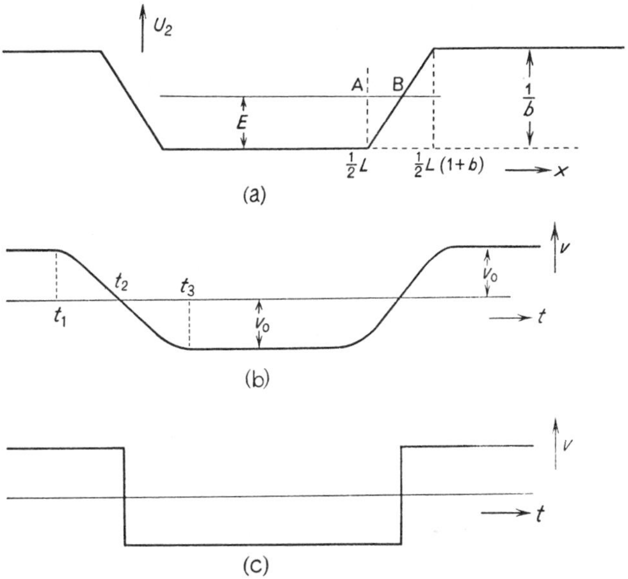

Fig. 7. a: The potential U_2 corresponding to two walls of finite height $1/b$;
b: Particle velocity v as a function of time for finite wall height;
c: Particle velocity v as a function of time for infinitely high walls.

its velocity has reached its original value, but in the opposite direction. In fig. 7b we have drawn the particle velocity as a function of time. The time t_1 corresponds to $x = \frac{1}{2}L$ when the velocity v is still equal to $v_0 = (2E/m)^{\frac{1}{2}}$. At t_2, $v = 0$ and $x = \frac{1}{2}L + \frac{1}{2}Eb^2L$ (we have assumed $E < 1/b$ so that $\frac{1}{2}L + \frac{1}{2}Eb^2L < \frac{1}{2}L(1+b)$). At time t_3 corresponding to a return to A $v = -v_0$ and again $x = \frac{1}{2}L$. The time τ spent in the 'wall region', $\frac{1}{2}L \leqq x \leqq \frac{1}{2}L(1+b)$, is given by the equation

$$\tau = 2(t_2 - t_1) = 2\int_{\frac{1}{2}L}^{\frac{1}{2}L(1+Eb^2)} \left[v_0^2 + \frac{2}{mb^2} - \frac{4x}{mb^2L}\right]^{-\frac{1}{2}} \mathrm{d}x,$$

or

$$\tau = mv_0 b^2 L.$$

We see that as $b \to 0$, $\tau \to 0$: the particle bounces off the wall, and its velocity is changed discontinuously, as shewn in fig. 7c. In that limit the movement of the particle is restricted to the range

$$-\tfrac{1}{2}L \leqq x \leqq \tfrac{1}{2}L, \tag{2.109}$$

that is, it is restricted to move inside a one-dimensional box with elastic walls.

The third potential leads to the following equations of motion,

$$m\ddot{x} = c\alpha(\alpha x+\beta y+\gamma z-d), \tag{2.110a}$$

$$m\ddot{y} = c\beta(\alpha x+\beta y+\gamma z-d), \tag{2.110b}$$

$$m\ddot{z} = c\gamma(\alpha x+\beta y+\gamma z-d); \tag{2.110c}$$

adding α times (2.110a), β times (2.110b), and γ times (2.110c), and solving the resultant differential equation, we find

$$\alpha x+\beta y+\gamma z-d = \sqrt{\frac{2E}{c}} \sin \sqrt{\frac{c}{m}}(t-t_0), \tag{2.111}$$

where E is the energy corresponding to the motion of the particle perpendicular to the plane

$$\alpha x+\beta y+\gamma z = d. \tag{2.112}$$

If now we take the limit $c \to \infty$, we find that the particle is constrained to move in the plane, whose equation is (2.112).

Considering the three cases together, we see that we can have constraints of the following kinds: (i) constraints which fix the distances apart of particles in the system (first case); (ii) constraints which require a particle to move on a given surface, or along a given curve (third case); (iii) constraints which require a particle — or a system of particles — to move in a restricted part of space (second case). The first two constraints can be expressed in the form of equations to be satisfied by the coordinates,

$$G_l(\boldsymbol{x}_1, \ldots, \boldsymbol{x}_i, \ldots, \boldsymbol{x}_N) = 0. \tag{2.113}$$

Such equations are called *kinematic relations*, or rather *holonomic kinematic relations*. The last restraint is expressed by an inequality,

$$\bar{G}_m(\boldsymbol{x}_1, \ldots, \boldsymbol{x}_i, \ldots, \boldsymbol{x}_N) \geqq 0, \tag{2.114}$$

and we talk about a non-holonomic constraint, just as the first two are called holonomic constraints. Another type of non-holonomic constraint is one which can only be expressed in differential form such as, for instance, the requirement of pure rolling of a sphere over a plane. In the following we shall consider only holonomic constraints, apart from a brief discussion in § 2.5 of the non-integrable constraints.

One way of dealing with constraints is to say that our system does not possess $3N$ degrees of freedom, but only s, given by the equation

$$s = 3N-p, \tag{2.115}$$

where p is the number of kinematic relations (2.113) which have to be satisfied. These kinematic relations will lead to forces in the system which see to it that they remain satisfied. If we call these forces $\boldsymbol{F}'_i$, we can split the total force acting on the ith particle into $\boldsymbol{F}'_i$ and the force $\boldsymbol{F}_i$ due to other sources, which leads to the equations of motion [compare (2.101)]

$$m_i\ddot{\boldsymbol{x}}_i = \boldsymbol{F}_i+\boldsymbol{F}'_i. \tag{2.116}$$

Our unknown quantities are the $\boldsymbol{x}_i$ and $\boldsymbol{F}'_i$, that is, $6N$ variables, while we have only $3N$ equations (2.116) and p equations (2.113). We must thus find some other equations. This can be done by using d'Alembert's principle.

2. D'ALEMBERT'S PRINCIPLE

Let us consider once again the constraints which arose from the limiting cases of potentials (2.102) and (2.104). Let us consider virtual displacements $\delta\boldsymbol{x}_i$ of the particles, where we define a virtual displacement as one which does not violate the kinematic relations, and let us consider the work done by the $\boldsymbol{F}'_i$ when such a virtual displacement takes place. In the two cases considered we shall shew that the virtual work done by the forces of constraint vanishes.

In the three-dimensional case (2.107) we are dealing with two particles, and the two forces $\boldsymbol{F}'_1$ and $\boldsymbol{F}'_2$ are equal and opposite (Newton's third law!) and directed along the line connecting the two particles,

$$\boldsymbol{F}'_1 = -\boldsymbol{F}'_2 = -k'(\boldsymbol{x}_1-\boldsymbol{x}_2), \tag{2.201}$$

where k' is a scalar. The constraint (2.106) for the three-dimensional case is of the form (2.108):

$$r_{12}^2 = (\boldsymbol{x}_1-\boldsymbol{x}_2 \cdot \boldsymbol{x}_1-\boldsymbol{x}_2) = l^2, \tag{2.202}$$

and virtual displacements $\delta \boldsymbol{x}_1$ and $\delta \boldsymbol{x}_2$ which do not violate the constraint (2.202) must satisfy the equation

$$(\boldsymbol{x}_1 - \boldsymbol{x}_2 \cdot \delta \boldsymbol{x}_1 - \delta \boldsymbol{x}_2) = 0. \tag{2.203}$$

The virtual work δW done by the forces of constraint is given by

$$\delta W = (\boldsymbol{F}_1' \cdot \delta \boldsymbol{x}_1) + (\boldsymbol{F}_2' \cdot \delta \boldsymbol{x}_2), \tag{2.204}$$

and using equations (2.201) and (2.203) we get

$$\begin{aligned} \delta W &= -k'(\boldsymbol{x}_1 - \boldsymbol{x}_2 \cdot \delta \boldsymbol{x}_1) + k'(\boldsymbol{x}_1 - \boldsymbol{x}_2 \cdot \delta \boldsymbol{x}_2) \\ &= -k'(\boldsymbol{x}_1 - \boldsymbol{x}_2 \cdot \delta \boldsymbol{x}_1 - \delta \boldsymbol{x}_2) = 0, \end{aligned} \tag{2.205}$$

which proves our statement for this case.

In the other case we write the constraint (2.112) in the form

$$(\boldsymbol{n} \cdot \boldsymbol{x}) = d, \tag{2.206}$$

where $\boldsymbol{n}$ is a unit vector with direction cosines α, β and γ. The force of constraint $\boldsymbol{F}'$ is directed perpendicular to the plane (2.206), or

$$\boldsymbol{F}' = k\boldsymbol{n}, \tag{2.207}$$

where k is a scalar, while the virtual displacement $\delta \boldsymbol{x}$ satisfies the condition

$$(\boldsymbol{n} \cdot \delta \boldsymbol{x}) = 0. \tag{2.208}$$

Combining (2.207) and (2.208) we find for the virtual work δW done by $\boldsymbol{F}'$,

$$\delta W = (\boldsymbol{F}' \cdot \delta \boldsymbol{x}) = k(\boldsymbol{n} \cdot \delta \boldsymbol{x}) = 0, \tag{2.209}$$

proving our statement also for this case.

We shall now define a *mechanical system* as a system for which the virtual work done by the forces of constraint vanishes. This is d'Alembert's principle: 'The virtual work done by the forces of constraint vanishes in any mechanical system.' In the following we shall only consider mechanical systems. The way we use d'Alembert's principle here is by defining our systems as those for which this principle is true, or, to put it differently, as those for which the forces of constraint are such that they cannot perform virtual work. Another way of introducing this principle, and one which is only slightly different, is by assuming it as a hypothesis which turns out to work. We may add that it will no longer hold if frictional effects are taken into account.

D'Alembert's principle can be expressed by the following equation

$$\sum_i (\boldsymbol{F}'_i \cdot \delta\boldsymbol{x}_i) = 0, \tag{2.210}$$

provided the $\delta\boldsymbol{x}_i$ satisfy the p equations

$$\sum_i (\delta\boldsymbol{x}_i \cdot \nabla_i G_l) = 0, \tag{2.211}$$

where the G_l are the functions entering into the kinematic relations (2.113). Equation (2.211) defines a virtual displacement, namely, a displacement where the $\boldsymbol{x}_i + \delta\boldsymbol{x}_i$ as well as the $\boldsymbol{x}_i$ satisfy each of the equations (2.113).

Using (2.210) and (2.211) we can now find another $3N-p$ relations which together with (2.116) and (2.113) will enable us to determine the motion of the particles completely. If there were no restrictions on the $\delta\boldsymbol{x}_i$, equation (2.210) would only be compatible with $\boldsymbol{F}'_i = 0$. This would lead to the original equations of motion ($m_i\ddot{\boldsymbol{x}}_i = \boldsymbol{F}_i$) for the case where there are no kinematic relations. The equations $\boldsymbol{F}'_i = 0$ would furnish us with the missing $3N$ conditions. If there are kinematic relations, the $\delta\boldsymbol{x}_i$ are not all independent, but they must satisfy (2.211). We can therefore only choose $3N-p$ of the $3N$ components of the $\delta\boldsymbol{x}_i$ arbitrarily, the last p components being determined by the conditions (2.211). We can try to eliminate p of these components from (2.210) by adding to it the p equations (2.211) each multiplied by a suitably chosen factor λ_l. The result would be

$$\sum_i (\delta\boldsymbol{x}_i \cdot \boldsymbol{F}'_i + \sum_l \lambda_l \nabla_i G_l) = 0. \tag{2.212}$$

We now choose the λ_l in such a way that the coefficients of the first p components $\delta\boldsymbol{x}_i$ vanish,

$$\boldsymbol{F}'_i + \sum_l \lambda_l \nabla_i G_l = 0. \tag{2.213}$$

Equation (2.213) is thus valid for p components and determines the λ_l. Substituting these values of λ_l into (2.212) we see that the sum over i, which originally meant a sum over $3N$ terms ($i = 1, \ldots, N$ and three terms in each scalar product) is reduced to a sum over $3N-p$ terms. However, the remaining $3N-p$ components of the $\delta\boldsymbol{x}_i$ are now independent and (2.212) can only be satisfied, if their coefficients vanish, that is, if (2.213) is also valid for the remaining $3N-p$ components. This equation is the one which enables us to solve, in principle, the equations of motion. We have $3N$ equations (2.116), p equations (2.113), and $3N$ equations (2.213) for $3N$ $\boldsymbol{x}_i$, $3N$ $\boldsymbol{F}'_i$, and p λ_l. The method used here is called the method of the undetermined multipliers, or of the Lagrangian multipliers. We shall return presently to the physical meaning of the Lagrangian multipliers λ_l.

From (2.213) and (2.116) we are led to the *Lagrangian equations of motion of the first kind,*

$$m_i \ddot{\boldsymbol{x}}_i = \boldsymbol{F}_i - \sum_{l=1}^{p} \lambda_l \boldsymbol{\nabla}_i G_l, \tag{2.214}$$

together with the p kinematic relations (2.113). We may note here that, if one speaks of the Lagrangian equations of motion, one practically always refers to those of the second kind which we shall encounter in the next section.

From (2.213) it is clear that the λ_l are intimately connected with the forces of constraint. We shall illustrate this by a few simple examples. The simplest is probably the constraint (2.206) which restricted the motion of a particle to a plane. In that case there is one kinematic relation, and the corresponding function G is given by the equation

$$G(\boldsymbol{x}) = (\boldsymbol{n} \cdot \boldsymbol{x}) - d \tag{2.215}$$

so that we get from (2.213)

$$\boldsymbol{F}' = -\lambda \boldsymbol{n}. \tag{2.216}$$

We see that in this case λ is the absolute magnitude of the force of constraint.

It is of interest to note that from (2.110) we get for the force component F_n perpendicular to the plane (2.112) in the case where c is still finite

$$F_n = m(\ddot{\boldsymbol{x}} \cdot \boldsymbol{n}) = c[(\boldsymbol{n} \cdot \boldsymbol{x}) - d]. \tag{2.217}$$

The multiplier λ is thus the limiting value of the expression on the right hand side of (2.217).

A similar situation arises when we consider Atwood's machine consisting of two masses m_1 and m_2 connected by a weightless string of fixed length which can move freely over a weightless pulley (fig. 8). The equation of constraint is now

$$G(z_1, z_2) \equiv z_1 + z_2 - l = 0. \tag{2.218}$$

From (2.213) we have now

$$F_1' = -\lambda, \qquad F_2' = -\lambda \tag{2.219}$$

and the equations of motion are

$$m_1 \ddot{z}_1 = -m_1 g - \lambda, \qquad m_2 \ddot{z}_2 = -m_2 g - \lambda, \tag{2.220}$$

where g is the gravitational acceleration. From these equations we see that $-\lambda$ is the tension in the string. The equations of motion can be solved, using

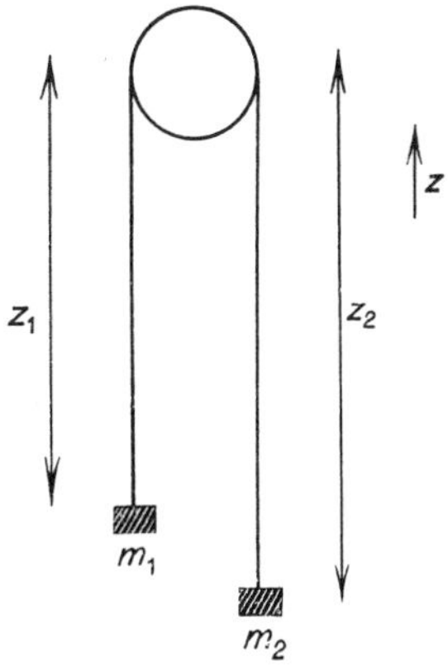

Fig. 8. Atwood's machine.

(2.218) and one finds that the heavier mass, say m_1, moves with a uniform downward acceleration a which is given by

$$a = \frac{m_1 - m_2}{m_1 + m_2} g, \tag{2.221}$$

while the tension is given by

$$-\lambda = \frac{2m_1 m_2}{m_1 + m_2} g. \tag{2.222}$$

The result (2.221) is, of course, intuitively correct: the driving force is equal to the difference between $m_1 g$ and $m_2 g$ while the inertia is provided by the sum of the two masses.

The last example will be that of a dumbbell molecule the centre of mass of which is constrained to move in a plane. The equations of constraint are in that case

$$\begin{aligned} G_1 &\equiv (x_1 - x_2)^2 + (y_1 - y_2)^2 + (z_1 - z_2)^2 - l^2 = 0, \\ G_2 &\equiv (m_1 \boldsymbol{x}_1 + m_2 \boldsymbol{x}_2 \cdot \boldsymbol{n}) - d = 0. \end{aligned} \tag{2.223}$$

From (2.213) we get

$$\begin{aligned} \boldsymbol{F}_1' &= -2\lambda_1(\boldsymbol{x}_1-\boldsymbol{x}_2)-\lambda_2 m_1\boldsymbol{n}, \\ \boldsymbol{F}_2' &= 2\lambda_1(\boldsymbol{x}_1-\boldsymbol{x}_2)-\lambda_2 m_2\boldsymbol{n}. \end{aligned} \tag{2.224}$$

From these equations we see that there are forces of constraint acting along the axis of the molecule. These forces are proportional to λ_1 and obey Newton's third law (action is reaction). There are also forces of constraint proportional to λ_2. If we construct their resultant acting upon the centre of mass of the molecule we get

$$\boldsymbol{F}'_{\text{c. o. m.}} = -\lambda_2(m_1+m_2)\,\boldsymbol{n}, \tag{2.225}$$

which is completely analogous to the force of constraint (2.216), but now refers to the centre of mass of the molecule instead of to a single particle.

Let us now return to the discussion of the general case. We can combine (2.210) and (2.116) to write d'Alembert's principle in the form

$$\sum_i (m_i\ddot{\boldsymbol{x}}_i-\boldsymbol{F}_i\cdot\delta\boldsymbol{x}_i) = 0, \tag{2.226}$$

where the $\delta\boldsymbol{x}_i$ satisfy equations (2.211). Consider now two possible orbits of the system, that is, two orbits for which the kinematic relations are satisfied and let them correspond to the values $\boldsymbol{x}_i(t)$ and $\boldsymbol{x}_i(t)+\delta\boldsymbol{x}_i(t)$ of the coordinates of the particles. If the two orbits are neighbouring orbits, we can treat the $\delta\boldsymbol{x}_i(t)$ as small quantities. For the variation $\delta\dot{\boldsymbol{x}}_i$ of the velocities $\dot{\boldsymbol{x}}_i$ we find

$$\delta\dot{\boldsymbol{x}}_i = \frac{\mathrm{d}}{\mathrm{d}t}(\boldsymbol{x}_i+\delta\boldsymbol{x}_i)-\frac{\mathrm{d}}{\mathrm{d}t}\boldsymbol{x}_i = \frac{\mathrm{d}}{\mathrm{d}t}\delta\boldsymbol{x}_i, \tag{2.227}$$

shewing the principle of variational calculus that differentiation and variation can be interchanged. If we consider the two orbits during a time interval t_1, t_2 we have from (2.226)

$$\int_{t_1}^{t_2}\sum_i (m_i\ddot{\boldsymbol{x}}_i-\boldsymbol{F}_i\cdot\delta\boldsymbol{x}_i)\mathrm{d}t = 0, \tag{2.228}$$

since the integrand vanishes at every moment. Using (2.227) and integrating the first term by parts we have

$$0 = \sum_i m_i(\dot{\boldsymbol{x}}_i\cdot\delta\boldsymbol{x}_i)\Big|_{t_1}^{t_2} - \int_{t_1}^{t_2}[\delta T+\sum_i(\boldsymbol{F}_i\cdot\delta\boldsymbol{x}_i)]\mathrm{d}t, \tag{2.229}$$

where T is the total kinetic energy

$$T = \tfrac{1}{2}\sum_i m_i\dot{\boldsymbol{x}}_i^2. \tag{2.230}$$

If we now consider such systems where the $\boldsymbol{F}_i$ can be derived from a potential U,

$$\boldsymbol{F}_i = -\nabla_i U, \tag{2.231}$$

so that

$$\sum_i (\boldsymbol{F}_i \cdot \delta \boldsymbol{x}_i) = -\delta U, \tag{2.232}$$

and such variations $\delta \boldsymbol{x}_i$ that

$$\delta \boldsymbol{x}_i = 0, \quad \text{at } t = t_1 \text{ and at } t = t_2, \tag{2.233}$$

we get from (2.229), (2.232), and (2.233),

$$\int_{t_1}^{t_2} (\delta T - \delta U) dt = \delta \int_{t_1}^{t_2} (T - U) dt = \delta \int_{t_1}^{t_2} L dt = 0, \tag{2.234}$$

where L is the *Lagrangian* defined by the equation

$$L = T - U. \tag{2.235}$$

Equation (2.234) is called *Hamilton's variational principle*; it is valid provided the end points of the orbit are not changed. It is equivalent to either d'Alembert's principle or the original equations of motion, provided the kinematic relations are satisfied by the orbits. The advantage of (2.234) over the two other statements is that it is independent of the choice of coordinates with which we describe the system.

3. LAGRANGE'S EQUATIONS OF MOTION

We have mentioned earlier that the kinematic relations will restrict the number of degrees of freedom of our system to $3N - p$ $(= s)$ and in many cases it is more convenient to introduce s independent variables which completely fix the state of the system than to continue using the N $\boldsymbol{x}_i$ ($3N$ Cartesian coordinates) together with the kinematic relations and the λ_l. It must be realised that in going over to *generalised coordinates* q_k ($k = 1, \ldots, s$), as those parameters are called, one is still dealing with mechanical systems for which d'Alembert's principle holds; this hypothesis is now, however, no longer clearly visible. The q_k will be functions of the $\boldsymbol{x}_i$, and conversely, while the *generalised velocities* $\dot{q}_k$ are related to the $\dot{\boldsymbol{x}}_i$ by the equations

$$\dot{\boldsymbol{x}}_i = \sum_k \frac{\partial \boldsymbol{x}_i}{\partial q_k} \dot{q}_k. \tag{2.301}$$

Substituting these equations into (2.230) for the kinetic energy, we have

$$T = \tfrac{1}{2}\sum a_{kl}\dot{q}_k\dot{q}_l, \tag{2.302}$$

where the a_{kl} are functions of the q_k which are given by the equation

$$a_{kl} = a_{lk} = \sum_i m_i\left(\frac{\partial \boldsymbol{x}_i}{\partial q_k}\cdot\frac{\partial \boldsymbol{x}_i}{\partial q_l}\right). \tag{2.303}$$

We see that T is a homogeneous quadratic function in the $\dot{q}_k$.

Up to now we have always assumed the potential energy U to be a function of the coordinates, $\boldsymbol{x}_i$, only; it will therefore now be a function of the q_k only, $U(q_k)$, and the Lagrangian will be a function of the q_k and the $\dot{q}_k$ (we shall briefly discuss velocity dependent potential energies later on in this chapter),

$$L = L(q_k, \dot{q}_k) = T(q_k, \dot{q}_k) - U(q_k). \tag{2.304}$$

We shall now apply Hamilton's variational principle to the Lagrangian (2.304). Its variation δL is given by the equation

$$\delta L = \sum_k \frac{\partial L}{\partial \dot{q}_k}\,\delta\dot{q}_k + \sum_k \frac{\partial L}{\partial q_k}\,\delta q_k, \tag{2.305}$$

while condition (2.233) is now changed to

$$\delta q_k = 0 \quad \text{at } t = t_1 \text{ and at } t = t_2. \tag{2.306}$$

From the variational principle (2.234) we get now

$$\begin{aligned} 0 = \int_{t_1}^{t_2} \delta L\,\mathrm{d}t &= \int_{t_1}^{t_2} \sum_k \frac{\partial L}{\partial \dot{q}_k}\,\delta\dot{q}_k\,\mathrm{d}t + \int_{t_1}^{t_2} \sum_k \frac{\partial L}{\partial q_k}\,\delta q_k\,\mathrm{d}t \\ &= \sum_k \frac{\partial L}{\partial \dot{q}_k}\,\delta q_k\Big|_{t_1}^{t_2} - \int_{t_1}^{t_2} \sum_k \frac{\mathrm{d}}{\mathrm{d}t}\frac{\partial L}{\partial \dot{q}_k}\,\delta q_k\,\mathrm{d}t + \int_{t_1}^{t_2} \sum_k \frac{\partial L}{\partial q_k}\,\delta q_k\,\mathrm{d}t \\ &= \int_{t_1}^{t_2} \sum_k \left[\frac{\partial L}{\partial q_k} - \frac{\mathrm{d}}{\mathrm{d}t}\frac{\partial L}{\partial \dot{q}_k}\right]\delta q_k\,\mathrm{d}t. \end{aligned} \tag{2.307}$$

We have used here the fact that the $\delta\dot{q}_k$ and the δq_k are not independent of one another: the $\delta\dot{q}_k$ are the time derivatives of the δq_k [compare (2.227)].

We now can reap the benefit of the introduction of the generalised coordinates. Since there are as many q_k as there are degrees of freedom, the δq_k are independent functions of the time, and (2.307) can only be satisfied, if the following s equations hold,

$$\frac{\mathrm{d}}{\mathrm{d}t}\frac{\partial L}{\partial \dot{q}_k} - \frac{\partial L}{\partial q_k} = 0, \qquad k = 1, 2, \ldots, s. \tag{2.308}$$

Equations (2.308) are called the Lagrangian equations of motion of the second kind, or usually just the *Lagrangian equations of motion.* As the derivation of (2.308) is independent of the choice of coordinates, on going over from one set of generalised coordinates, q_k, to another set, q'_k, we will find the transformed equations

$$\frac{\mathrm{d}}{\mathrm{d}t}\frac{\partial L}{\partial \dot{q}'_k} - \frac{\partial L}{\partial q'_k} = 0. \tag{2.309}$$

Before discussing the Lagrangian equations of motion we introduce the *generalised momenta*, p_k, by the equation

$$p_k = \frac{\partial L}{\partial \dot{q}_k}. \tag{2.310}$$

In the case of Cartesian coordinates the p_k are just the components of the linear momenta of the different particles. Using the p_k we can write (2.308) in the form

$$\dot{p}_k = \frac{\partial L}{\partial q_k}. \tag{2.311}$$

Let us consider the application of the Lagrangian equations of motion to some physical systems. First of all we note that for a system without constraints we can use the Cartesian coordinates $\boldsymbol{x}_i$ as generalised coordinates. In that case we have

$$L = \tfrac{1}{2}\sum_i m_i(\dot{x}_i^2 + \dot{y}_i^2 + \dot{z}_i^2) - U(\boldsymbol{x}_1, \boldsymbol{x}_2, \ldots), \tag{2.312}$$

so that

$$\frac{\partial L}{\partial \dot{x}_i} = m_i\dot{x}_i, \quad \frac{\partial L}{\partial \dot{y}_i} = m_i\dot{y}_i, \quad \frac{\partial L}{\partial \dot{z}_i} = m_i\dot{z}_i, \quad \frac{\partial L}{\partial x_i} = -\frac{\partial U}{\partial x_i}, \ldots,$$

and (2.308) reduce to the Newtonian equations of motion (1.302).

As our next example we consider a single particle in a spherically symmetric potential, $U(r)$. We choose for our generalised coordinates $q_1 = r$, $q_2 = \theta$, and $q_3 = \varphi$, where r, θ, and φ are spherical polars. The kinetic energy is now given by the equation

$$T = \tfrac{1}{2}m(\dot{r}^2 + r^2\dot{\theta}^2 + r^2 \sin^2\theta\, \dot{\varphi}^2), \tag{2.313}$$

so that we have for the Lagrangian

$$L = \tfrac{1}{2}m(\dot{r}^2 + r^2\dot{\theta}^2 + r^2 \sin^2\theta\, \dot{\varphi}^2) - U(r). \tag{2.314}$$

The generalised momenta are in this case

$$p_r = \frac{\partial L}{\partial \dot{r}} = m\dot{r}, \tag{2.315}$$

$$p_\theta = \frac{\partial L}{\partial \dot{\theta}} = mr^2\dot{\theta}, \tag{2.316}$$

$$p_\varphi = \frac{\partial L}{\partial \dot{\varphi}} = mr^2 \sin^2\theta\, \dot{\varphi}. \tag{2.317}$$

One sees immediately that p_r is the radial momentum, that is, the linear momentum component in the direction of the radius vector. The momentum p_φ is the component of the total angular momentum around the polar axis. If the spherical polars r, θ, φ are related in the usual way to Cartesian coordinates,

$$x = r \sin\theta \cos\varphi, \qquad y = r \sin\theta \sin\varphi, \qquad z = r \cos\theta, \tag{2.318}$$

p_φ is equal to the z-component of the angular momentum $\boldsymbol{M}$ given by (1.209) as can be seen by substitution.

The square of the length of the angular momentum vector M^2 can be expressed in terms of p_θ and p_φ, as follows,

$$M^2 = p_\theta^2 + \frac{p_\varphi^2}{\sin^2\theta}. \tag{2.319}$$

From the Lagrangian equations of motion we get

$$\dot{p}_\varphi = 0, \tag{2.320}$$

$$\dot{p}_\theta = \frac{\partial L}{\partial \theta}. \tag{2.321}$$

The last equation leads to

$$\dot{p}_\theta = \frac{p_\varphi^2 \cot\theta}{mr^2 \sin^2\theta}, \tag{2.322}$$

and combining (2.319), (2.320), and (2.322) we find that $\dot{M} = 0$, or, M^2 = constant. If we choose the polar axis in such a way that it cuts the orbit somewhere, we see that in the point of intersection where $\theta = 0$ and r and $\dot{\varphi}$ are finite, p_φ will vanish, and since p_φ is a constant [from (2.320)] $p_\varphi = 0$ everywhere, and p_θ is now the total angular momentum. Equation (2.316) is then the same as (1.214) and the last of the equations of motion,

$$\dot{p}_r = \frac{\partial L}{\partial r}, \tag{2.323}$$

reduces to (1.227) when we use (2.314), (2.315) and $p_\varphi = 0$.

The next problem to be considered is Atwood's machine (see fig. 8). Instead of two coordinates z_1 and z_2 with the constraint (2.218) we introduce one generalised coordinate q ($= z_1$) so that $z_2 = l - q$. The potential energy U is given by the equation

$$U = -m_1 g q - m_2 g (l-q), \tag{2.324}$$

while the kinetic energy is given by

$$T = \tfrac{1}{2}(m_1+m_2)\dot{q}^2. \tag{2.325}$$

The Lagrangian equation of motion (2.308) leads to

$$(m_1+m_2)\ddot{q} = (m_1-m_2)g, \tag{2.326}$$

in accordance with (2.221). We do not, however, obtain an expression for the tension in the string, but we can evaluate it, of course, by considering the forces on one of the masses.

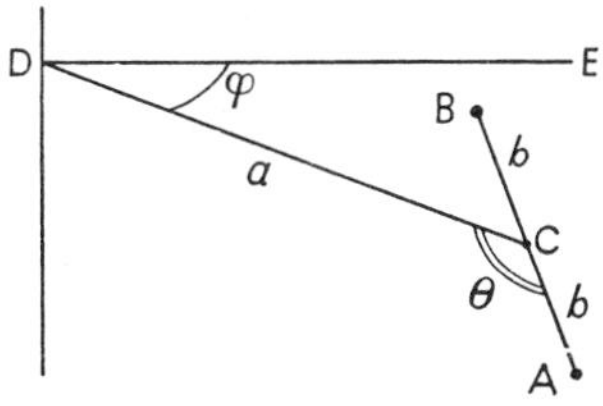

Fig. 9. The Thomson-Tait pendulum. The rod AB can move freely in a vertical plane through DC and this rod can itself move freely in the horizontal plane CDE.

Our last example is the so-called Thomson-Tait pendulum (fig. 9) which consists of two equal masses m fixed at the end of a weightless rod (AB) of length $2b$. The middle of this rod C is attached to the end of a weightless rod (CD) of length a. The rod CD is mounted in such a way that it can move freely in a horizontal plane, while AB is mounted so that it can rotate freely in the vertical plane through CD. There are two degrees of freedom, and we choose as our generalised coordinates the angle φ between CD and a fixed direction DE in the given horizontal plane and the angle θ between AB and CD (see fig. 9). As the two masses are the same, the potential energy in the gravitational field (assumed to be uniform) vanishes. The kinetic energy can be obtained from the expression

$$T = \tfrac{1}{2}m(\dot{x}_1^2+\dot{y}_1^2+\dot{z}_1^2+\dot{x}_2^2+\dot{y}_2^2+\dot{z}_2^2) \tag{2.327}$$

using the following expressions for x_1, y_1, z_1, x_2, y_2, and z_2,

$$\begin{aligned} x_1 &= a\cos\varphi + b\cos\theta\cos\varphi, & x_2 &= a\cos\varphi - b\cos\theta\cos\varphi,\\ y_1 &= a\sin\varphi + b\cos\theta\sin\varphi, & y_2 &= a\sin\varphi - b\cos\theta\sin\varphi,\\ z_1 &= b\sin\theta & z_2 &= -b\sin\theta. \end{aligned} \tag{2.328}$$

The final result for the kinetic energy — and thus for the Lagrangian — is

$$\begin{aligned} L = T &= \tfrac{1}{2}m[b^2\dot{\theta}^2+(a+b\cos\theta)^2\dot{\varphi}^2]+\tfrac{1}{2}m[b^2\dot{\theta}^2+(a-b\cos\theta)^2\dot{\varphi}^2]\\ &= mb^2\dot{\theta}^2+m(a^2+b^2\cos^2\theta)\dot{\varphi}^2. \end{aligned} \tag{2.329}$$

From (2.308) we get the following equations of motion

$$\frac{\mathrm{d}}{\mathrm{d}t}\frac{\partial L}{\partial\dot{\varphi}} = \dot{p}_\varphi = 0, \quad \text{or,} \quad p_\varphi = \text{constant}, \tag{2.330}$$

$$2mb^2\ddot{\theta}+2mb^2\cos\theta\sin\theta\dot{\varphi}^2 = 0. \tag{2.331}$$

Combining (2.330), (2.331), and the equation for p_φ,

$$p_\varphi = 2m(a^2+b^2\cos^2\theta)\dot{\varphi}, \tag{2.332}$$

we get

$$\ddot{\theta} + \frac{p_\varphi^2\sin\theta\cos\theta}{4m^2(a^2+b^2\cos^2\theta)^2} = 0, \tag{2.333}$$

from which θ can be obtained by a double quadrature. We shall return to (2.333) in the next section.

4. CYCLIC COORDINATES

Often there are coordinates which do not occur in the Lagrangian themselves, although their time derivatives do. In the preceding section we met several instances, for instance, the polar angle φ in the Lagrangian (2.314), or the angle φ in expression (2.329). Such a coordinate will be called a *cyclic* or *ignorable* coordinate. The first term refers to the fact that such coordinates are very often angles — as in the two cases referred to —, while the reason for the second term will become clear presently. Let q_s be an ignorable coordinate. In that case we find from (2.308) and (2.310),

$$\frac{\mathrm{d}}{\mathrm{d}t}\frac{\partial L}{\partial \dot{q}_s} = \frac{\partial L}{\partial q_s} = 0 = \frac{\mathrm{d}p_s}{\mathrm{d}t}, \tag{2.401}$$

or

$$p_s = \text{constant}. \tag{2.402}$$

Equation (2.402) can be used to ignore the degree of freedom connected with q_s as follows. Equation (2.402) is a relation between $\dot{q}_1, \dot{q}_2, \ldots, \dot{q}_{s-1}, \dot{q}_s, q_1, q_2, \ldots, q_{s-1}$, and (the constant) p_s. We can solve this equation to express $\dot{q}_s$ in terms of the other variables,

$$\dot{q}_s = \dot{q}_s(\dot{q}_1, \dot{q}_2, \ldots, \dot{q}_{s-1}, q_1, \ldots, q_{s-1}; p_s), \tag{2.403}$$

where once again we remind ourselves that p_s is a constant. We now introduce the so-called Routh function R by the equation

$$R = L - p_s\dot{q}_s, \tag{2.404}$$

and using (2.402) and (2.403) we find that

$$R = R(\dot{q}_1, \ldots, \dot{q}_{s-1}, q_1, \ldots, q_{s-1}, p_s). \tag{2.405}$$

To find the equations of motion in terms of R we write

$$\delta L = \sum_{k=1}^{s} \frac{\partial L}{\partial \dot{q}_k}\,\delta\dot{q}_k + \sum_{k=1}^{s} \frac{\partial L}{\partial q_k}\,\delta q_k$$

$$= \sum_{k=1}^{s-1} \frac{\partial L}{\partial \dot{q}_k}\,\delta\dot{q}_k + \sum_{k=1}^{s-1} \frac{\partial L}{\partial q_k}\,\delta q_k + p_s\delta\dot{q}_s + \frac{\partial L}{\partial q_s}\,\delta q_s,$$

whence it follows that

$$\delta R = \delta(L - p_s\dot{q}_s) = \delta L - p_s\delta\dot{q}_s - \dot{q}_s\delta p_s$$

$$= \sum_{k=1}^{s-1} p_k\delta\dot{q}_k + \sum_{k=1}^{s-1} \dot{p}_k\delta q_k - \dot{q}_s\delta p_s, \tag{2.406}$$

where we have used (2.310) for the p_k and (2.401). From (2.405) we find

$$p_k = \frac{\partial R}{\partial \dot{q}_k}, \qquad \dot{p}_k = \frac{\partial R}{\partial q_k}, \qquad k = 1, \ldots, s-1, \tag{2.407}$$

or,

$$\frac{\mathrm{d}}{\mathrm{d}t}\frac{\partial R}{\partial \dot{q}_k} - \frac{\partial R}{\partial q_k} = 0, \qquad k = 1, \ldots, s-1. \tag{2.408}$$

Equations (2.408) can be solved, and once all p_k and q_k are found for $k = 1, \ldots, s-1$, we can use the equation

$$\dot{q}_s = -\frac{\partial R}{\partial p_s} \tag{2.409}$$

to give us q_s as a function of time. We note that the (2.408) are of the same form as the Lagrangian equations (2.308), but there is one less equation; also (2.407) are of the same form as (2.310) and (2.311).

If there are more cyclic coordinates, we can ignore all of them simultaneously by using the Routh function

$$R = L - \sum_i p_i \dot{q}_i, \tag{2.410}$$

where the summation is over all ignorable degrees of freedom.

We shall illustrate the use of a Routh function by shewing how the centre of mass degrees of freedom can be ignored. Consider a system of N particles with coordinates $\boldsymbol{x}_i$ ($i = 1, \ldots, N$) and a potential energy which depends on the relative positions of the particles only,

$$U = U(\boldsymbol{x}_i - \boldsymbol{x}_j). \tag{2.411}$$

We introduce centre of mass coordinates,

$$M\boldsymbol{X} = \sum_i m_i \boldsymbol{x}_i, \qquad M = \sum_i m_i, \tag{2.412}$$

and relative coordinates

$$\boldsymbol{x}_i' = \boldsymbol{x}_i - \boldsymbol{X}, \tag{2.413}$$

of which there are only $3N-3$, since they satisfy the relation

$$\sum_i m_i \boldsymbol{x}_i' = 0, \tag{2.414}$$

as follows easily from (2.412) and (2.413).

The kinetic energy of the system is given by the equation

$$T = \sum_i \tfrac{1}{2}m_i(\dot{\boldsymbol{x}}_i \cdot \dot{\boldsymbol{x}}_i) = \tfrac{1}{2}M(\dot{\boldsymbol{X}} \cdot \dot{\boldsymbol{X}}) + \sum_i m_i(\dot{\boldsymbol{x}}_i' \cdot \dot{\boldsymbol{X}}) + \tfrac{1}{2}\sum_i m_i(\dot{\boldsymbol{x}}_i' \cdot \dot{\boldsymbol{x}}_i')$$
$$= T_{\text{com}} + T_{\text{rel}}, \tag{2.415}$$

where we have used (2.414), and where T_{com} and T_{rel} are respectively the kinetic energy relating to the motion of the centre of mass, and the kinetic energy relating to the relative motion of the particles,

$$T_{\text{com}} = \tfrac{1}{2}M(\dot{\boldsymbol{X}} \cdot \dot{\boldsymbol{X}}), \qquad T_{\text{rel}} = \tfrac{1}{2}\sum_i m_i(\dot{\boldsymbol{x}}_i' \cdot \dot{\boldsymbol{x}}_i'). \tag{2.416}$$

From (2.411) it follows that the potential energy is a function of the $\boldsymbol{x}_i'$ only so that the Lagrangian

$$L = T - U = T_{\text{com}} + T_{\text{rel}} - U \tag{2.417}$$

does not contain the centre of mass coordinates, which are thus ignorable. We introduce the Routh function

$$R = L - (\boldsymbol{P}_{\text{com}} \cdot \dot{\boldsymbol{X}}), \tag{2.418}$$

where

$$\boldsymbol{P}_{\text{com}} = \frac{\partial L}{\partial \dot{\boldsymbol{X}}} = M\dot{\boldsymbol{X}} \tag{2.419}$$

is the vector of the total linear momentum. The function R is now equal to

$$R = -T_{\text{com}} + T_{\text{rel}} - U, \tag{2.420}$$

and the equations (2.408) lead to the equations of motion for the relative motion, while (2.409) describes the uniform centre of mass motion.

To conclude this section we wish to discuss briefly a topic which has lost most of its actuality but which was of great interest in the heyday of classical mechanics. Consider again for a moment the case of the Thomson-Tait pendulum discussed in the previous section. The angle φ was ignorable. We found that we could eliminate φ and obtain the equation of motion (2.333) for the other coordinate θ. This same equation could also have been obtained by introducing a Routh function and ignoring φ in the manner just described. If we look at (2.333) we see that although there was no potential energy, this equation is of the same form as an equation of motion for a one-dimensional problem with a potential energy. By such cases Hertz was led to the conclusion that there was never any potential energy: it only occurs when we do not consider the total system. This point of view is nowadays no longer adhered to; indeed, one might argue the other way round as we saw at the beginning of this chapter when we showed how the kinematic relations can be considered to occur through limiting processes, starting from suitably chosen potential energies.

One also spoke of *hidden masses* in this connexion; these hidden masses were assumed to produce through the kinematic relations the pseudo-potential-energies, such as the one in (2.333). We shall consider a very simple case where, indeed, a mass which is hidden produces such a pseudo-

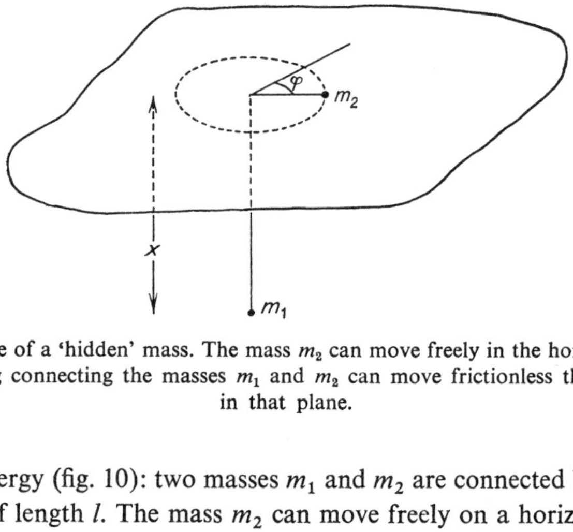

Fig. 10. A case of a 'hidden' mass. The mass m_2 can move freely in the horizontal plane, and the string connecting the masses m_1 and m_2 can move frictionless through a hole in that plane.

potential-energy (fig. 10): two masses m_1 and m_2 are connected by a weightless string of length l. The mass m_2 can move freely on a horizontal plane, while the string can move frictionless through a hole in the plane so that the mass m_1 moves vertically. There are two degrees of freedom and we take as our generalised coordinates x and φ (see fig. 10). The Lagrangian for this system is

$$L = \tfrac{1}{2}m_1\dot{x}^2 + \tfrac{1}{2}m_2\dot{x}^2 + \tfrac{1}{2}m_2(l-x)^2\dot{\varphi}^2 + m_1 g x. \tag{2.421}$$

As φ is ignorable we can ignore it and use the Routh function

$$R = L - p_\varphi\dot{\varphi} = \tfrac{1}{2}(m_1+m_2)\dot{x}^2 - \frac{p_\varphi^2}{2m_2(l-x)^2} + m_1 g x. \tag{2.422}$$

The system is reduced to one of one degree of freedom only, and a 'pseudo-potential', U', has appeared:

$$U'(x) = \frac{p_\varphi^2}{2m_2(l-x)^2}. \tag{2.423}$$

5. NON-HOLONOMIC CONSTRAINTS; VELOCITY-DEPENDENT POTENTIALS

Although we do not wish to enter into a detailed discussion of non-holonomic constraints, there is one interesting class of non-holonomic constraints which we wish to mention briefly. This is the class of non-integrable constraints. As an example of a physical system where such constraints occur we may mention the hoop (see, for instance, Nr. 4 of the Problem Section at the end of this book). If x and y are the coordinates of the point where the hoop rests on the ground, and if θ is an angle measuring the amount by which the hoop has turned, the condition for pure rolling is

$$\delta x^2 + \delta y^2 = R^2 \delta\theta^2,$$

where R is the radius of the hoop.

Let us consider a system where apart from p constraints of the kind (2.113) there are r non-integrable constraints of the kind

$$\sum_k a_k^{(j)} \delta q_k = 0, \qquad j = 1, \ldots, r. \tag{2.501}$$

In (2.501) we have already introduced generalised coordinates q_k, that is, we have taken the holonomic constraints into account. Equation (2.307) still holds, so that we have

$$\int_{t_1}^{t_2} \sum_k \left[\frac{\partial L}{\partial q_k} - \frac{\mathrm{d}}{\mathrm{d}t}\frac{\partial L}{\partial \dot{q}_k}\right] \delta q_k \mathrm{d}t = 0, \tag{2.502a}$$

provided

$$\delta q_k = 0 \quad \text{at } t = t_1 \text{ and at } t = t_2, \tag{2.502b}$$

but the δq_k are no longer independent, since they must satisfy at any time the relations (2.501). The answer to this difficulty is once again the method of the Lagrangian multipliers which now leads to

$$\int_{t_1}^{t_2} \sum_k \left[\frac{\partial L}{\partial q_k} - \frac{\mathrm{d}}{\mathrm{d}t}\frac{\partial L}{\partial \dot{q}_k} + \sum_{j=1}^{r} \lambda_j a_k^{(j)}\right] \delta q_k \mathrm{d}t = 0. \tag{2.503}$$

Now we can treat the δq_k as independent variations (compare the discussion in § 2.2) and we get the equations of motion

$$\frac{\mathrm{d}}{\mathrm{d}t}\frac{\partial L}{\partial \dot{q}_k} - \frac{\partial L}{\partial q_k} - \sum_{j=1}^{r} \lambda_j a_k^{(j)} = 0, \tag{2.504}$$

a form which is intermediate between the Lagrangian equations of motion of the first kind and those of the second kind. The λ_j are once again related to the forces of constraint and (2.504) must be combined with (2.501) to determine both the q_k and the λ_j as functions of the time. An example of such an intermediate case is given by problem 4 of the problem section at the end of the book.

Although we have assumed throughout that the potential energy U is a function of the q_k only, this was nowhere used in deriving (2.308) which are, indeed, also valid for the case where U depends on the $\dot{q}_k$ as well as on the q_k. In that case we can write these equations in the form

$$\frac{\mathrm{d}}{\mathrm{d}t}\frac{\partial T}{\partial \dot{q}_k} - \frac{\mathrm{d}}{\mathrm{d}t}\frac{\partial U}{\partial \dot{q}_k} - \frac{\partial T}{\partial q_k} + \frac{\partial U}{\partial q_k} = 0. \tag{2.505}$$

An important example of such a case is that of the motion of a point particle in an electromagnetic field †. Let $\boldsymbol{E}$ and $\boldsymbol{B}$ be the electrical field strength and the magnetic induction which can be derived from a scalar potential ϕ and a vector potential $\boldsymbol{A}$,

$$\boldsymbol{E} = -\nabla\phi - \frac{\partial \boldsymbol{A}}{\partial t}, \qquad \boldsymbol{B} = [\nabla \wedge \boldsymbol{A}]. \tag{2.506}$$

The equations of motion can be derived from (2.505) if we use for the potential energy the expression

$$U = e\phi - e(\boldsymbol{A} \cdot \dot{\boldsymbol{x}}). \tag{2.507}$$

Indeed, using the relation

$$\frac{\mathrm{d}\boldsymbol{A}}{\mathrm{d}t} = \frac{\partial \boldsymbol{A}}{\partial t} + (\dot{\boldsymbol{x}} \cdot \nabla)\boldsymbol{A}, \tag{2.508}$$

we get from (2.505)

$$m\ddot{\boldsymbol{x}} = -e\nabla\phi - e\frac{\partial \boldsymbol{A}}{\partial t} + e[\dot{\boldsymbol{x}} \wedge [\nabla \wedge \boldsymbol{A}]] = \boldsymbol{F}_{\mathrm{Lor}}, \tag{2.509}$$

† We use rationalised mks units for electromagnetic quantities.

where $\boldsymbol{F}_{\text{Lor}}$ is the Lorentz force,

$$\boldsymbol{F}_{\text{Lor}} = e\{\boldsymbol{E}+[\dot{\boldsymbol{x}} \wedge \boldsymbol{B}]\}. \tag{2.510}$$

6. CONSERVATION LAWS

The fact that any momentum corresponding to an ignorable coordinate is a constant of motion [see (2.402)] leads us to a general theorem that if the Lagrangian is invariant under a group of transformations which contains infinitesimal transformations, it is possible to find conservation laws. We shall in this section consider three conservation laws of this kind.

Let us consider a transformation

$$\boldsymbol{x}_i \to \boldsymbol{x}_i+\delta\boldsymbol{x}_i, \quad \text{or,} \quad q_k \to q_k+\delta q_k. \tag{2.601}$$

Under such a transformation the change in the Lagrangian δL will be given by the equation

$$\delta L = \sum_i (\nabla_i L \cdot \delta\boldsymbol{x}_i), \quad \text{or,} \quad \delta L = \sum_k \frac{\partial L}{\partial q_k}\,\delta q_k. \tag{2.602}$$

The first form applies to a description of the system by Cartesian coordinates and the second to one by generalised coordinates q_k. If the group of transformations mentioned at the beginning of this section can be described by a change in just one suitably chosen, generalised coordinate, q_r say, we have that for any arbitrary change in q_r, δq_r, δL will vanish, and thus

$$\frac{\partial L}{\partial q_r} = 0, \tag{2.603}$$

from which follows that

$$p_r = \text{constant}. \tag{2.604}$$

In § 2.4 we saw that in a system where the potentials present depend only on the relative positions of the particles, the momentum corresponding to the centre of mass motion is a constant of motion. The appropriate q_r would thus be the three components of the total linear momentum. To see this we note that for a system of this kind the Lagrangian is invariant under all translations, that is, it is invariant under all transformations

$$\boldsymbol{x}_i \to \boldsymbol{x}_i+\boldsymbol{\varepsilon}, \tag{2.605}$$

where $\boldsymbol{\varepsilon}$ is an arbitrary vector. From (2.602) it then follows that

$$\delta L = 0 = \sum_i (\boldsymbol{\nabla}_i L \cdot \boldsymbol{\varepsilon}), \tag{2.606}$$

or,

$$\sum_i \boldsymbol{\nabla}_i L = 0,$$

since $\boldsymbol{\varepsilon}$ is arbitrary. Using the Lagrangian equations of motion (2.308) we get

$$\sum_i \frac{\mathrm{d}}{\mathrm{d}t} \frac{\partial L}{\partial \dot{\boldsymbol{x}}_i} = 0, \tag{2.607}$$

or,

$$\sum_i \frac{\partial L}{\partial \dot{\boldsymbol{x}}_i} = \sum_i m_i \dot{\boldsymbol{x}}_i = \boldsymbol{P}_{\text{com}} = \text{constant}. \tag{2.608}$$

We see thus that the conservation of total linear momentum follows from the fact that the Lagrangian is invariant under all translations.

Consider now the case where the Lagrangian is invariant under rotation around an axis. Let this axis be parallel to the unit vector $\boldsymbol{n}$ and let $\delta\varphi$ be the angle over which the system is rotated. The q_r of (2.603) is thus φ and we would expect the corresponding p_φ to be a constant. This p_φ is the angular momentum around the given axis. We can see this most easily by remembering that a rotation around $\boldsymbol{n}$ over an angle $\delta\varphi$ corresponds to the transformation

$$\boldsymbol{x}_i \to \boldsymbol{x}_i + \delta\varphi\, [\boldsymbol{n} \wedge \boldsymbol{x}_i], \tag{2.609}$$

and (2.602) corresponds to

$$\delta L = 0 = \sum_i \delta\varphi (\boldsymbol{\nabla}_i L \cdot [\boldsymbol{n} \wedge \boldsymbol{x}_i]), \tag{2.610}$$

which can be rewritten as

$$\delta\varphi \sum_i (\boldsymbol{n} \cdot [\boldsymbol{x}_i \wedge \boldsymbol{\nabla}_i L]) = \delta\varphi \left(\boldsymbol{n} \cdot \frac{\mathrm{d}}{\mathrm{d}t} \sum_i [\boldsymbol{x}_i \wedge m_i \dot{\boldsymbol{x}}_i] \right) = 0, \tag{2.611}$$

or

$$(\boldsymbol{n} \cdot \sum_i [\boldsymbol{x}_i \wedge m_i \dot{\boldsymbol{x}}_i]) = \text{constant} = M_{\boldsymbol{n}}, \tag{2.612}$$

where $M_{\boldsymbol{n}}$ is the component of the total angular momentum (1.309) along $\boldsymbol{n}$.

It follows thus that if the Lagrangian is invariant under all rotations, the total angular momentum of the system is conserved.

Let us finally consider the case where the Lagrangian is invariant with respect to a change in the time coordinate, that is, where L does not depend explicitly on the time, or

$$\frac{\partial L}{\partial t} = 0. \tag{2.613}$$

We find in that case

$$\begin{aligned}\frac{dL}{dt} &= \sum_k \frac{\partial L}{\partial q_k}\dot{q}_k + \sum_k \frac{\partial L}{\partial \dot{q}_k}\ddot{q}_k + \frac{\partial L}{\partial t} \\ &= \sum_k \frac{d}{dt}\frac{\partial L}{\partial \dot{q}_k}\dot{q}_k + \sum_k \frac{\partial L}{\partial \dot{q}_k}\frac{d\dot{q}_k}{dt} = \sum_k \frac{d}{dt}\left(\dot{q}_k \frac{\partial L}{\partial \dot{q}_k}\right),\end{aligned} \tag{2.614}$$

where we have used the Lagrangian equations of motion and (2.613). In the case in which we have mainly been interested where the potential energy U depends on the q_k only so that the $\dot{q}_k$ only appear in the kinetic energy T which is a homogeneous quadratic expression in the $\dot{q}_k$, we have

$$\sum_k \dot{q}_k \frac{\partial L}{\partial \dot{q}_k} = \sum_k \dot{q}_k \frac{\partial T}{\partial \dot{q}_k} = 2T, \tag{2.615}$$

and from (2.614) we get

$$\frac{d}{dt}\left(L - \sum_k \dot{q}_k \frac{\partial L}{\partial \dot{q}_k}\right) = 0 = \frac{d}{dt}(T - U - 2T), \tag{2.616}$$

whence it follows that

$$E = T + U = \text{constant}, \tag{2.617}$$

where E is the total energy: If the Lagrangian is invariant under a change in the time, conservation of energy follows. This is really a special case of the first conservation law if we treat Cartesian coordinates on a par with the time coordinate (compare § 5.4).

CHAPTER 3

SMALL VIBRATIONS

The general theory of small vibrations around an equilibrium position is developed in some detail and it is shewn how one can obtain the normal coordinates. The theory is illustrated by considering the small vibrations of a double pendulum, the molecular vibrations of some simple molecules, and the normal vibrations of a one-dimensional crystal. The cases discussed are those of diatomic and of non-linear and linear triatomic molecules of the A_2B type. Finally, a simple case of vibrations around an equilibrium motion is considered.

1. THE THEORY OF SMALL VIBRATIONS

We shall consider in this section first of all the equilibrium states of mechanical systems, that is, systems whose motion can be described by the Lagrangian equations of motion,

$$\frac{\mathrm{d}}{\mathrm{d}t}\frac{\partial L}{\partial \dot{q}_k} - \frac{\partial L}{\partial q_k} = 0, \tag{3.101}$$

and we shall once again assume that the potential energy U does not contain the generalised velocities so that the Lagrangian L is given by the equation

$$L(q_k, \dot{q}_k) = T(q_k, \dot{q}_k) - U(q_k), \tag{3.102}$$

where T is the kinetic energy.

An equilibrium state of the system is defined as a state for which the generalised coordinates q_k have values $q_k^{(0)}$ which are such that, if $\dot{q}_k = 0$ while $q_k = q_k^{(0)}$, all higher time derivatives of the q_k vanish also:

$$q_k = q_k^{(0)} \quad \text{and} \quad \dot{q}_k = 0 \rightarrow \ddot{q}_k = 0, \quad \dddot{q}_k = 0, \quad \ddddot{q}_k = 0, \ldots \tag{3.103}$$

This means that if all the particles in the system are at rest at the positions corresponding to $q_k = q_k^{(0)}$ they will stay at rest, or, $q_k = q_k^{(0)}$ together with $\dot{q}_k = 0$ at $t = t_0$, will lead to $q_k = q_k^{(0)}$ at any later time.

We saw in the previous chapter, (2.302), that the kinetic energy can be written in the form

$$T = \tfrac{1}{2} \sum_{k,l} a_{kl} \dot{q}_k \dot{q}_l, \tag{3.104}$$

where the $a_{kl} (= a_{lk})$ are in general functions of the q_k. Using expression (3.104) for T we get from (3.101)

$$\sum_l a_{kl} \ddot{q}_l + \sum_{l,m} \frac{\partial a_{kl}}{\partial q_m} \dot{q}_l \dot{q}_m - \tfrac{1}{2} \sum_{l,m} \frac{\partial a_{lm}}{\partial q_k} \dot{q}_l \dot{q}_m + \frac{\partial U}{\partial q_k} = 0. \tag{3.105}$$

If we introduce the so-called Christoffel symbol $\left\{ {lm \atop k} \right\}$,

$$\left\{ {lm \atop k} \right\} = \tfrac{1}{2} \left[\frac{\partial a_{lk}}{\partial q_m} + \frac{\partial a_{mk}}{\partial q_l} - \frac{\partial a_{lm}}{\partial q_k} \right], \tag{3.106}$$

we can write (3.105) in the form

$$\sum_l a_{kl} \ddot{q}_l + \sum_{l,m} \left\{ {lm \atop k} \right\} \dot{q}_l \dot{q}_m = F_k^{(\mathrm{gen})}, \tag{3.107}$$

where we have introduced the generalised forces $F_k^{(\mathrm{gen})}$,

$$F_k^{(\mathrm{gen})} = - \frac{\partial U}{\partial q_k}. \tag{3.108}$$

The Christoffel symbols play an important role in Riemannian geometry. For a discussion of their importance in the formulation of classical mechanics we refer to H. C. Corben and P. Stehle, Classical Mechanics (J. Wiley and Sons, New York, 1950).

From the equilibrium condition (3.103) and equation (3.105) it follows that a necessary condition for equilibrium is that

$$\frac{\partial U}{\partial q_k} = 0, \quad k = 1, \ldots, s \quad \text{when} \quad q_l = q_l^{(0)} \quad (l = 1, \ldots, s). \tag{3.109}$$

This means that U is extremum, or rather, has a stationary value, at equilibrium. There are several possibilities, some of which are illustrated in fig. 11 for the case where $s = 2$. Fig. 11a corresponds to an absolute minimum and the equilibrium state is a stable one. Fig. 11d corresponds to an absolute maximum and fig. 11b to a saddle-point; in both cases the equilibrium state is an unstable one. Fig. 11c finally corresponds to an indifferent equilibrium.

We shall now consider small departures from equilibrium. To simplify our equations we shall shift the origin in q-space in such a way that for the equilibrium state considered (it must be remembered that there may be several sets of q_k-values satisfying the equilibrium condition) all $q_k^{(0)}$ are equal to zero. This means that the values of the q_k will be small when we consider small departures from equilibrium and that we can use power series expansions in terms of the q_k, and only consider the first few terms.

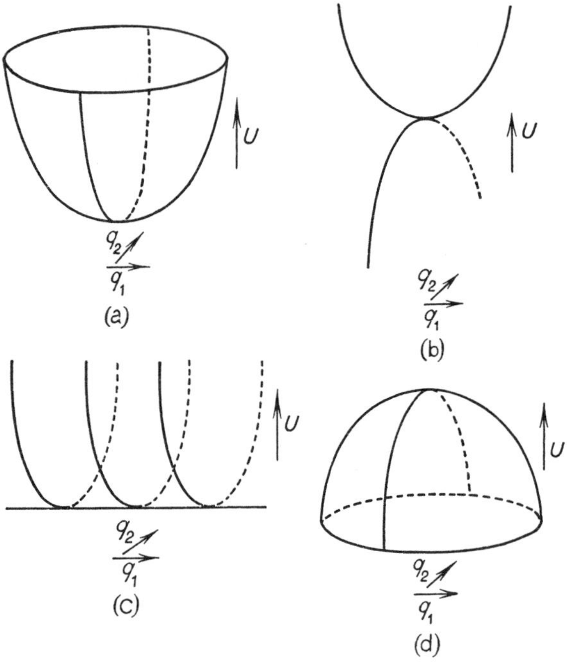

Fig. 11. Possible shapes of a two-dimensional potential energy as a function of the generalised coordinates for different equilibrium states.
a: Stable equilibrium; b: unstable equilibrium; c: indifferent equilibrium; d: unstable equilibrium.

In this way we get for the kinetic and potential energies

$$U(q) = U(0) + \sum_k \left(\frac{\partial U}{\partial q_k}\right)_0 q_k + \tfrac{1}{2} \sum_{k,l} \left(\frac{\partial^2 U}{\partial q_k \partial q_l}\right)_0 q_k q_l + \ldots, \tag{3.110}$$

$$T(q, \dot{q}) = \tfrac{1}{2} \sum_{k,l} \left[(a_{kl})_0 + \sum_m \left(\frac{\partial a_{kl}}{\partial q_m}\right)_0 q_m + \ldots \right] \dot{q}_k \dot{q}_l. \tag{3.111}$$

Using (3.109), putting $U(0) = 0$, and dropping all terms of third order or higher in U and of first order or higher in T, we get for the Lagrangian

$$L = T - U = \tfrac{1}{2}\sum_{k,l} c_{kl}\dot{q}_k\dot{q}_l - \tfrac{1}{2}\sum_{k,l} b_{kl}q_kq_l, \tag{3.112}$$

where

$$c_{kl} = (a_{kl})_0 = c_{lk}, \qquad b_{kl} = \left(\frac{\partial^2 U}{\partial q_k \partial q_l}\right)_0 = b_{lk}. \tag{3.113}$$

The Lagrangian equations of motion (3.101) now lead to the basic equations of the theory of small vibrations:

$$\sum_l c_{kl}\ddot{q}_l + \sum_l b_{kl}q_l = 0, \qquad k = 1, \ldots, s. \tag{3.114}$$

To solve these equations we shall look for eigenvibrations of the system, that is, such solutions that all q_k oscillate with the same frequency. This is equivalent — as we shall see presently — to looking for another set of coordinates, Q_m, such that in those coordinates, the equations of motion take on the form

$$\ddot{Q}_m + \lambda_m Q_m = 0, \qquad m = 1, \ldots, s. \tag{3.115}$$

The Q_m are obtained from the q_k by a linear transformation which means that if for an eigenvibration the q_k satisfy the equations

$$q_k = A_k e^{i\omega t}, \tag{3.116}$$

where ω is the same for all q_k, the amplitudes A_k will satisfy certain relations.

To see all this we start from (3.114). Multiplying each equation by a constant α_k and summing over k we get

$$\sum_{k,l} c_{kl}\alpha_k\ddot{q}_l + \sum_{k,l} b_{kl}\alpha_k q_l = 0. \tag{3.117}$$

If this equation is to be the same as (3.115) where Q_m is a linear combination of the q_k,

$$Q_m = \sum_k \beta_k^{(m)} q_k, \tag{3.118}$$

we find that the α_k and the $\beta_k^{(m)}$ must satisfy the equations

$$\sum_l b_{kl}\alpha_l = \lambda_m \beta_k^{(m)}, \qquad \sum_l c_{kl}\alpha_l = \beta_k^{(m)}, \tag{3.119}$$

or,

$$\sum_l (b_{kl} - \lambda_m c_{kl})\alpha_l = 0, \qquad \sum_l c_{kl}\alpha_l = \beta_k^{(m)}. \tag{3.120}$$

The first of these equations possesses a non-trivial solution for the α_l, that is, a solution different from the trivial solution $\alpha_l = 0$, $l = 1, \ldots, s$, only if λ_m is one of the roots of the determinantal equation

$$\begin{vmatrix} b_{11}-\lambda c_{11} & \ldots\ldots & b_{1s}-\lambda c_{1s} \\ \vdots & \ddots & \vdots \\ b_{s1}-\lambda c_{s1} & \ldots\ldots & b_{ss}-\lambda c_{ss} \end{vmatrix} = 0. \tag{3.121}$$

If λ_m is one such root we can determine from (3.120) a set of ratios for the α_k and thus also for the $\beta_k^{(m)}$ so that we have found a linear combination Q_m of the q_k satisfying (3.115). We note first of all that it follows from the theory of linear equations that α_k will be proportional to the co-determinant of the k-th element of any one of the rows of the determinant of (3.121). It thus follows that the set of ratios we found for the α_k depends on the value λ_m. Equation (3.121) possesses s roots and there will thus be s sets of α_k ($\alpha_k^{(m)}$, $m = 1, \ldots, s$) and hence s sets of $\beta_k^{(m)}$ and thus s Q_m. The procedure outlined here is straightforward as long as the roots λ_m are non-degenerate, that is, as long as no two roots are equal, but the situation is more complicated if there are multiple roots. We do not wish to discuss here how one should proceed, if multiple roots are present [(see, for instance, H. A. Kramers, Quantum Theory (North Holland Publishing Company, Amsterdam, 1957, § 37)]. If there are no multiple roots we have s different values λ_m and we find s different sets of α_k and s different Q_m. We remark here that also in the case of multiple roots it is possible to find such a set of different Q_m for which the $\alpha_k^{(m)}$ satisfy, moreover, (3.131). In the rest of this section we shall for the sake of simplicity assume that all the roots of (3.121) are different.

First of all we shall prove that all λ_m are real. To do this we rewrite (3.119) in the form

$$\sum_l b_{kl}\alpha_l^{(m)} = \lambda_m \sum_l c_{kl}\alpha_l^{(m)}, \tag{3.122}$$

where we have indicated by the superscript (m) that the α_k are different for different values of λ_m. Multiplying (3.122) by $\alpha_k^{(m)*}$ and summing over k we get

$$\sum_{k,l} b_{kl}\alpha_k^{(m)*}\alpha_l^{(m)} = \lambda_m \sum_{k,l} c_{kl}\alpha_k^{(m)*}\alpha_l^{(m)}. \tag{3.123}$$

Subtracting from this equation its complex conjugate we get

$$(\lambda_m - \lambda_m^*) \sum_{k,l} c_{k,l} \alpha_k^{(m)*} \alpha_l^{(m)} = 0. \tag{3.124}$$

As $c_{kl} = c_{lk}$ and as the c_{kl} are real, the sum over k and l is real. Moreover, this sum can be written in the form

$$\sum_{k,l} c_{kl} \alpha_k^{(m)*} \alpha_l^{(m)} = \sum_{k,l} c_{kl} a_k^{(m)} a_l^{(m)} + \sum_{k,l} c_{kl} b_k^{(m)} b_l^{(m)}, \tag{3.125}$$

where $a_k^{(m)}$ and $b_k^{(m)}$ are the real and imaginary parts of $\alpha_k^{(m)}$. It follows from the physical meaning of $\frac{1}{2} \sum_{k,l} c_{kl} \dot{q}_k \dot{q}_l$ as a kinetic energy that the two sums on the right hand side of (3.125) are positive definite. It thus follows that we must have from (3.124)

$$\lambda_m = \lambda_m^*, \tag{3.126}$$

or, that all the λ_m are real. We see that this is a consequence of the fact that the kinetic energy is positive definite.

If the λ_m are real, it follows from the fact that the $\alpha_k^{(m)}$ are proportional to the co-determinants in the determinant in (3.121) that we can choose all $\alpha_k^{(m)}$ also to be real. (As we can multiply any set $\alpha_k^{(m)}$ for a given value of m by a common factor, which could be complex, we cannot prove that all $\alpha_k^{(m)}$ *are* real, but only that we can choose them to be real.) We shall assume that we have made such a choice for each set $\alpha_k^{(m)}$ (m fixed). We shall choose it in such a way that

$$\sum_{k,l} c_{kl} \alpha_k^{(m)} \alpha_l^{(m)} = 1. \tag{3.127}$$

This equation we shall call the *normalisation* of the $\alpha_k^{(m)}$. If we multiply (3.122) by $\alpha_k^{(n)}$ and sum over k we get

$$\sum_{k,l} b_{kl} \alpha_l^{(m)} \alpha_k^{(n)} = \lambda_m \sum_{k,l} c_{kl} \alpha_l^{(m)} \alpha_k^{(n)}. \tag{3.128}$$

Subtracting from this equation the same equation, but with m and n interchanged, we get

$$(\lambda_m - \lambda_n) \sum_{k,l} c_{kl} \alpha_k^{(m)} \alpha_l^{(n)} = 0, \tag{3.129}$$

from which follows that if $m \neq n$

$$\sum_{k,l} c_{kl} \alpha_k^{(m)} \alpha_l^{(n)} = 0, \tag{3.130}$$

and we can combine (3.127) and (3.130) into the following *orthonormalisation conditions,*

$$\sum_{k,\,l} c_{kl}\alpha_k^{(m)}\alpha_l^{(n)} = \delta_{mn}, \tag{3.131}$$

where δ_{mn} is the Kronecker symbol,

$$\delta_{mn} = 0, \quad m \neq n; \qquad \delta_{mn} = 1, \quad m = n. \tag{3.132}$$

If we multiply (3.131) by $\alpha_r^{(n)}$ and sum over n we get

$$\sum_k \Big[\sum_{l,\,n} c_{kl}\alpha_l^{(n)}\alpha_r^{(n)}\Big]\alpha_k^{(m)} = \alpha_r^{(m)}, \tag{3.133}$$

from which it follows that

$$\sum_{l,\,n} c_{kl}\alpha_l^{(n)}\alpha_r^{(n)} = \delta_{kr}. \tag{3.134}$$

From (3.123) it follows that

$$\lambda_m = \frac{\tfrac{1}{2}\sum b_{kl}\alpha_k^{(m)}\alpha_l^{(m)}}{\tfrac{1}{2}\sum c_{kl}\alpha_k^{(m)}\alpha_l^{(m)}}. \tag{3.135}$$

In this expression the denominator is the value of the kinetic energy when the $\dot{q}_k$ are equal to $\alpha_k^{(m)}$. If the $\alpha_k^{(m)}$ are normalised according to (3.131), the denominator reduces to $\frac{1}{2}$. The enumerator is the value of the potential energy when the q_k are equal to $\alpha_k^{(m)}$. If we are in the neighbourhood of a stable equilibrium (see fig. 11a) this expression will be positive definite, and all λ_m will be positive. If, however, we are in the neighbourhood of an equilibrium such as the one given by fig. 11d, the enumerator will be negative definite, and all λ_m will be negative. In the case corresponding to fig. 11b we would expect both positive and negative λ_m, while an indifferent equilibrium will lead to at least one vanishing λ_m. In the next sections we shall meet with all these possibilities.

Let us now return to the Q_m for which we have from (3.118) and (3.119)

$$Q_m = \sum_{k,\,l} c_{kl}\alpha_l^{(m)}q_k. \tag{3.136}$$

Multiplying this equation by $\alpha_r^{(m)}$, summing over m, and using (3.134), we get

$$\sum_m \alpha_r^{(m)}Q_m = \sum_{k,\,l,\,m} c_{kl}\alpha_l^{(m)}\alpha_r^{(m)}q_k = q_r. \tag{3.137}$$

Equations (3.136) and (3.137) together gives us the transformation from the q_k to the Q_m and the other way round.

Expressed in the Q_m and the $\dot{Q}_m$ we get for the kinetic and potential energies

$$T = \tfrac{1}{2}\sum_{k,l} c_{kl}\dot{q}_k\dot{q}_l = \tfrac{1}{2}\sum_{k,l,m,n} c_{kl}\alpha_k^{(m)}\alpha_l^{(n)}\dot{Q}_m\dot{Q}_n = \tfrac{1}{2}\sum_m \dot{Q}_m^2, \tag{3.138}$$

$$U = \tfrac{1}{2}\sum_{k,l} b_{kl}q_kq_l = \tfrac{1}{2}\sum_{k,l,m,n} b_{kl}\alpha_k^{(m)}\alpha_l^{(n)}Q_mQ_n$$

$$= \tfrac{1}{2}\sum_{m,n}\lambda_m\sum_{k,l} c_{kl}\alpha_k^{(m)}\alpha_l^{(n)}Q_mQ_n = \tfrac{1}{2}\sum_m \lambda_m Q_m^2, \tag{3.139}$$

where we have used (3.131) and (3.128).

We note that the transformation (3.137) reduces both T and U to a sum of squares, or, brings them both on principal axes.

The equations of motion reduce, of course, to (3.115) with the solutions

$$Q_m = Q_m^{(0)}\mathrm{e}^{\pm i\omega_m t}, \tag{3.140}$$

where

$$\omega_m = \sqrt{\lambda_m}. \tag{3.141}$$

We note that positive values of λ_m correspond to real ω_m and thus to real oscillations, while negative λ_m lead to purely imaginary ω_m and thus to monotonically increasing or decreasing amplitudes. We see now the connexion between the shape of the potential energy in the neighbourhood of the equilibrium position (fig. 11) and the stability of the equilibrium.

The Q_m are called the *normal coordinates* and each of them corresponds to one of the *normal frequencies*. If only one of them is different from zero, we get for the q_k expressions of the form (3.116) where the common frequency ω is equal to ω_m, while the amplitudes A_k are proportional to the $\alpha_k^{(m)}$. In general the q_k are a linear combination of the different Q_m, given by (3.137).

2. THE DOUBLE PENDULUM

We shall illustrate the theory of the preceding section by a few simple examples. The first example is that of the so-called double pendulum (fig. 12). This pendulum consists of two masses M and m; the first one is at a fixed distance a from the point of suspension P, while the second one is at a fixed distance b from the first mass. We shall only consider motion in a plane so that there are two degrees of freedom. Moreover, to simplify our calculations we shall put $a = b$. The angles φ and ψ (fig. 12) will be used as generalised coordinates. The potential energy U is given by the equation

$$U = C - Mga\cos\varphi - mga(\cos\varphi + \cos\psi), \tag{3.201}$$

where C is a constant fixing the zero of potential energy and g the gravitational acceleration.

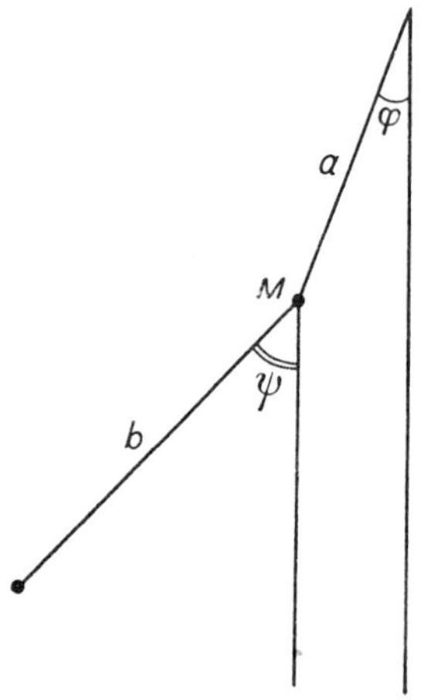

Fig. 12. The double pendulum.

The equilibrium positions are determined from the equations

$$\frac{\partial U}{\partial \varphi} = \mu g a \sin\varphi = 0, \qquad \frac{\partial U}{\partial \psi} = mga\sin\psi = 0, \tag{3.202}$$

where μ is the sum of the two masses,

$$\mu = m + M. \tag{3.203}$$

Equations (3.202) possess four sets of solutions:

$$\begin{aligned} &\text{(i)}\quad \varphi_{\text{eq}} = \psi_{\text{eq}} = 0; &&\text{(ii)}\quad \varphi_{\text{eq}} = 0,\quad \psi_{\text{eq}} = \pi; \\ &\text{(iii)}\quad \varphi_{\text{eq}} = \pi,\quad \psi_{\text{eq}} = 0; &&\text{(iv)}\quad \varphi_{\text{eq}} = \psi_{\text{eq}} = \pi. \end{aligned} \tag{3.204}$$

We see immediately that solution (i) corresponds to the stable equilibrium, and the potential energy in the neighbourhood of this equilibrium position behaves as in fig. 11a. Solutions (ii) and (iii) correspond to fig. 11b and solution (iv) to fig. 11d.

If we write $q_1 = \varphi - \varphi_{\text{eq}}$, $q_2 = \psi - \psi_{\text{eq}}$, we get for the potential energy in the small amplitude approximation

$$\begin{aligned} U-U_{eq} &= \tfrac{1}{2}ga[\mu q_1^2+mq_2^2], \quad &(i);\\ &= \tfrac{1}{2}ga[\mu q_1^2-mq_2^2], \quad &(ii);\\ &= \tfrac{1}{2}ga[-\mu q_1^2+mq_2^2], \quad &(iii);\\ &= \tfrac{1}{2}ga[-\mu q_1^2-mq_2^2], \quad &(iv). \end{aligned} \tag{3.205}$$

The kinetic energy T is easily seen to be given by the expression

$$T = \tfrac{1}{2}\mu a^2\dot{\varphi}^2+\tfrac{1}{2}ma^2\dot{\psi}^2+ma^2\dot{\varphi}\dot{\psi}\cos(\varphi-\psi),$$

or,

$$\begin{aligned} T &= \tfrac{1}{2}a^2[\mu\dot{q}_1^2+2m\dot{q}_1\dot{q}_2+m\dot{q}_2^2], \quad &(i), (iv);\\ &= \tfrac{1}{2}a^2[\mu\dot{q}_1^2-2m\dot{q}_1\dot{q}_2+m\dot{q}_2^2], \quad &(ii), (iii). \end{aligned} \tag{3.206}$$

Equation (3.121) now leads to

$$\begin{vmatrix} \lambda\mu a^2-\mu ga & \lambda ma^2\\ \lambda ma^2 & \lambda ma^2-mga \end{vmatrix} = 0, \ (i); \qquad \begin{vmatrix} \lambda\mu a^2-\mu ga & -\lambda ma^2\\ -\lambda ma^2 & \lambda ma^2+mga \end{vmatrix} = 0, \ (ii);$$
$$\begin{vmatrix} \lambda\mu a^2+\mu ga & -\lambda ma^2\\ -\lambda ma^2 & \lambda ma^2-mga \end{vmatrix} = 0, \ (iii); \qquad \begin{vmatrix} \lambda\mu a^2+\mu ga & \lambda ma^2\\ \lambda ma^2 & \lambda ma^2+mga \end{vmatrix} = 0, \ (iv). \tag{3.207}$$

These equations lead to the following equations for λ:

$$\begin{aligned} &(i): &Ma^2\lambda^2-2\mu ga\lambda+\mu g^2 &= 0;\\ &(ii), (iii): &Ma^2\lambda^2-\mu g^2 &= 0;\\ &(iv): &Ma^2\lambda^2+2\mu ga\lambda+\mu g^2 &= 0. \end{aligned} \tag{3.208}$$

The roots of the eigenvalue equation (3.207) are thus

$$\begin{aligned} &(i): &\lambda_{1,2} &= (\mu g/Ma)[1\pm\sqrt{m/\mu}];\\ &(ii), (iii): &\lambda_{1,2} &= \pm(g/a)\sqrt{\mu/M};\\ &(iv): &\lambda_{1,2} &= -(\mu g/Ma)[1\pm\sqrt{m/\mu}], \end{aligned} \tag{3.209}$$

where we have used (3.203).

We see that in case (i) both λ are positive (as $m < \mu$), in cases (ii) and (iii) we have one positive and one negative λ, while in case (iv) both λ are negative. This is in accordance with what was discussed in connexion with (3.135). It is instructive to consider the normal coordinates for the cases (i), (ii), and (iii). We find these from (3.136) and (3.131). The second of these two equations gives us the normalisation of the $\alpha_k^{(m)}$; if this normalisation is not taken into account, one finds that the Q_m do not necessarily lead to both a kinetic energy of the form (3.138) and a potential energy of the form (3.139). It is very simple to get the ratios of the $\alpha_k^{(m)}$, and slightly tedious to get their normalisation. The final result is

$$
\text{(i)}: \quad \begin{aligned}
Q_1 &= a\sqrt{\tfrac{1}{2}-\tfrac{1}{2}\sqrt{\frac{m}{\mu}}}\,[\sqrt{\mu}\cdot q_1-\sqrt{m}\cdot q_2],\\
Q_2 &= a\sqrt{\tfrac{1}{2}+\tfrac{1}{2}\sqrt{\frac{m}{\mu}}}\,[\sqrt{\mu}\cdot q_1+\sqrt{m}\cdot q_2];
\end{aligned}
$$

$$
\text{(ii)}: \quad \begin{aligned}
Q_1 &= a\sqrt{\tfrac{1}{2}\mu+\tfrac{1}{2}\sqrt{M\mu}}\left[\left(\sqrt{\frac{M}{m}}-\sqrt{\frac{\mu}{m}}\right)q_1+\sqrt{\frac{m}{\mu}}\,q_2\right],\\
Q_2 &= a\sqrt{\tfrac{1}{2}\mu-\tfrac{1}{2}\sqrt{M\mu}}\left[\left(\sqrt{\frac{M}{m}}+\sqrt{\frac{\mu}{m}}\right)q_1-\sqrt{\frac{m}{\mu}}\,q_2\right];
\end{aligned} \tag{3.210}
$$

$$
\text{(iii)}: \quad \begin{aligned}
Q_1 &= a\sqrt{\tfrac{1}{2}\mu-\tfrac{1}{2}\sqrt{M\mu}}\left[\left(\sqrt{\frac{M}{m}}+\sqrt{\frac{\mu}{m}}\right)q_1-\sqrt{\frac{m}{\mu}}\,q_2\right],\\
Q_2 &= a\sqrt{\tfrac{1}{2}\mu+\tfrac{1}{2}\sqrt{M\mu}}\left[\left(\sqrt{\frac{M}{m}}-\sqrt{\frac{\mu}{m}}\right)q_1+\sqrt{\frac{m}{\mu}}\,q_2\right].
\end{aligned}
$$

To see what kind of motion corresponds to the normal modes, let us first of all consider case (i). If the system is in the first mode, Q_2 will be equal to zero which means that q_1 and q_2 have opposite signs and that the amplitude of q_2 is larger than that of q_1 by a factor $\sqrt{\mu/m}$. This means that this mode looks something like the one depicted in fig. 13a. Similarly we get for the second mode a motion like the one of fig. 13b. In cases (ii) and (iii) there is one stable mode (corresponding to the positive λ_1) and one unstable mode (for λ_2 which is negative). They are illustrated in figs. 13c—13f. We may draw attention to the fact that the stable mode in case (ii), for instance, involves a motion of both M and m. One might at first sight have guessed that the mass M would stay at rest. However, if M is finite there would be a reaction on M, if m were moving and a stable mode only occurs when both move (in opposite directions) in such a way that the potential energy actually increases.

It is of interest to note that if $M \to \infty$, the stable mode, indeed, corresponds to a motion of m only.

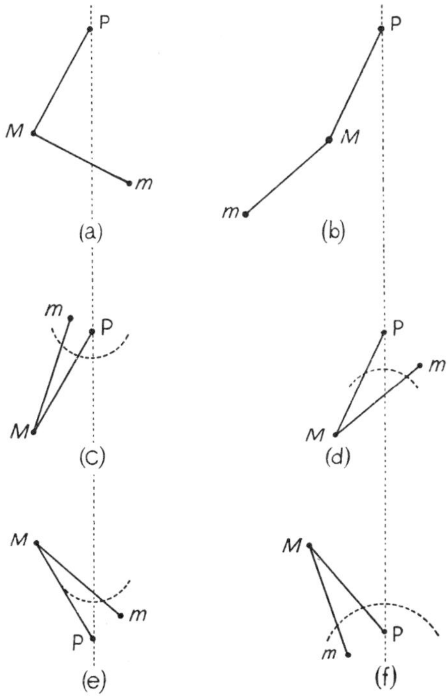

Fig. 13. Normal modes of the double pendulum; P: point of suspension.
a: First mode for case (i) where the potential energy is an absolute minimum;
b: Second mode for case (i) where the potential energy is an absolute minimum;
c: First (stable) mode for case (ii) where the potential energy has a saddle point;
d: Second (unstable) mode for case (ii) where the potential energy has a saddle point;
e: First (stable) mode for case (iii) where the potential energy has a saddle point;
f: Second (unstable) mode for case (iii) where the potential energy has a saddle point;
In Figures 13c, d, e, and f the path described by the centre of mass of the two masses m and M is given as a dotted curve.

In the limit $M \to 0$, we find for case (i) that for the first mode $\lambda_1 \to \infty$: the mass m stays vertically under the point of suspension P ($q_1 = -q_2$ in this case) while the mass M oscillates with an infinite frequency. For the second mode we find $q_1 = q_2$ and $\lambda_2 = g/2a$: the system swings as if we had a simple pendulum of length $2a$!

3. MOLECULAR VIBRATIONS

The theory of small vibrations is of importance in the study of molecular vibrations. In the present section we shall consider in some detail the cases of a diatomic molecule, such as HCl, of a non-linear triatomic molecule with a symmetrical equilibrium configuration, such as H_2O, and of a linear triatomic molecule, such as CO_2. For most discussions we shall assume that the interactions between the atoms in the molecules are additive and that they can be described by an interatomic potential of the general form depicted in fig. 14.

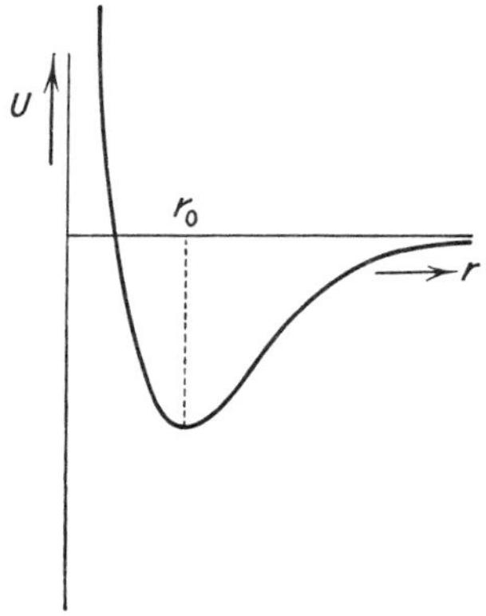

Fig. 14. The interatomic potential U as a function of the interatomic distance r; r_0: equilibrium distance.

Let us first of all consider the case of a diatomic molecule, which we shall treat as a dumbbell molecule, that is, a system of two masses, m_1 and m_2, interacting through a central potential $U(r)$. The Lagrangian of this system is given by the equation

$$L = \tfrac{1}{2}m_1(\dot{x}_1^2+\dot{y}_1^2+\dot{z}_1^2)+\tfrac{1}{2}m_2(\dot{x}_2^2+\dot{y}_2^2+\dot{z}_2^2)+U(r_{12}), \qquad (3.301)$$

where $r_{12} = \sqrt{(x_1-x_2)^2+(y_1-y_2)^2+(z_1-z_2)^2}$, and where x_1, y_1, z_1, x_2, y_2, and z_2 are the Cartesian coordinates of the two particles. We shall introduce as generalised coordinates the centre of mass coordinates X, Y, and Z, and spherical polar coordinates r, θ, and ϕ for the relative coordinates:

$$\begin{aligned} MX &= m_1x_1+m_2x_2, & MY &= m_1y_1+m_2y_2,\\ MZ &= m_1z_1+m_2z_2, & M &= m_1+m_2;\\ r\sin\theta\cos\phi &= x_1-x_2, & r\sin\theta\sin\phi &= y_1-y_2,\\ & & r\cos\theta = z_1-z_2. \end{aligned} \qquad (3.302)$$

The Lagrangian is in these coordinates

$$L = \tfrac{1}{2}M(\dot{X}^2+\dot{Y}^2+\dot{Z}^2)+\tfrac{1}{2}\mu(\dot{r}^2+r^2\dot{\theta}^2+r^2\sin^2\theta\,\dot{\phi}^2)-U(r), \tag{3.303}$$

where μ is the reduced mass,

$$\mu = m_1 m_2/M. \tag{3.304}$$

The equilibrium position is given by

$$\frac{\partial U}{\partial r} = 0, \quad \text{or,} \quad r = r_0 \qquad \text{(see fig. 14)}, \tag{3.305}$$

and we can choose the other five coordinates arbitrarily. Let us assume that we have chosen them to be equal to X_0, Y_0, Z_0, θ_0, and ϕ_0. The determinantal equation for the λ_m is now of the form

$$\begin{vmatrix} \lambda M & 0 & 0 & 0 & 0 & 0 \\ 0 & \lambda M & 0 & 0 & 0 & 0 \\ 0 & 0 & \lambda M & 0 & 0 & 0 \\ 0 & 0 & 0 & \lambda\mu - b & 0 & 0 \\ 0 & 0 & 0 & 0 & \lambda\mu r_0^2 & 0 \\ 0 & 0 & 0 & 0 & 0 & \lambda\mu r_0^2\sin^2\theta_0 \end{vmatrix} = 0, \tag{3.306}$$

where

$$b = \left(\frac{\mathrm{d}^2 U}{\mathrm{d}r^2}\right)_{r=r_0} \tag{3.307}$$

By standard methods — which are slightly complicated by the five-fold degeneracy of the root $\lambda = 0$ — we find the following values for the λ_m and for the corresponding Q_m [or rather: we find that we can choose the Q_1, Q_2, Q_3, Q_5, and Q_6 in the way given by (3.308)]:

$$\begin{aligned} \lambda_1 &= 0, & Q_1 &= X\sqrt{M}; \\ \lambda_2 &= 0, & Q_2 &= Y\sqrt{M}; \\ \lambda_3 &= 0, & Q_3 &= Z\sqrt{M}; \\ \lambda_4 &= b/\mu, & Q_4 &= r\sqrt{\mu}; \\ \lambda_5 &= 0, & Q_5 &= r_0\sqrt{\mu}\,\theta; \\ \lambda_6 &= 0, & Q_6 &= r_0\sqrt{\mu}\,\sin\theta_0\,\phi. \end{aligned} \tag{3.308}$$

Of the six normal coordinates, five are cyclic: Q_1, Q_2, Q_3, Q_5, and Q_6. Of those Q_1, Q_2, and Q_3 correspond to translations, while Q_5 and Q_6

correspond to rotations. The only non-cyclic coordinate is Q_4 corresponding to a vibration along the molecular axis.

It is of some interest to reduce this problem first to one in two dimensions and then to a one-dimensional one. This can be done by first ignoring Z and ϕ, and then ignoring Y and θ, in both cases by using the appropriate Routh function in the way described in § 2.4. Originally there were six degrees of freedom of which five (three translational and two rotational ones) were cyclic. In the two-dimensional case we drop in the Lagrangian (3.303) the terms $\frac{1}{2}M\dot{Z}^2$ and $\frac{1}{2}\mu r^2 \sin^2\theta\, \dot{\phi}^2$. There are now four degrees of freedom, corresponding to Q_1, Q_2, Q_4, and Q_5 of which three (two translational: Q_1 and Q_2, and one rotational: Q_5) are cyclic. The one-dimensional case, which is obtained by dropping also the tems $\frac{1}{2}M\dot{Y}^2$ and $\frac{1}{2}\mu r^2\dot{\theta}^2$, has two degrees of freedom (Q_1 and Q_4) of which Q_1 is a translational, cyclic one. In this case it is very easy to discard the cyclic degrees of freedom, but this is not always the case. In general, for instance, the elimination of the centre of mass motion of a many-particle system leads to equations which are much more complicated than the original ones.

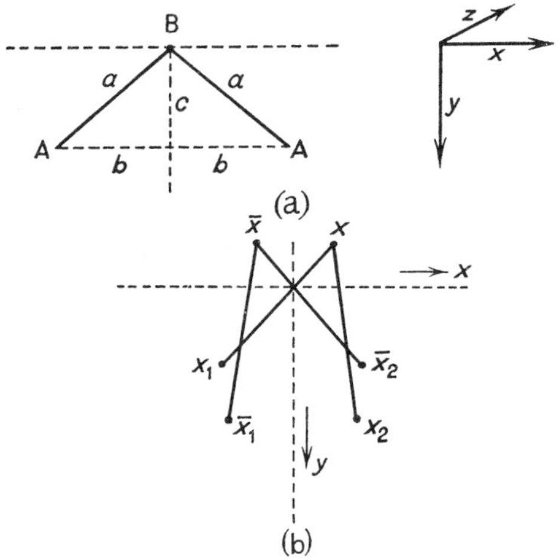

Fig. 15. a: Equilibrium configuration of an A_2B-molecule; b: Two non-equilibrium configurations of an A_2B-molecule with the same potential energy.

We shall now consider triatomic molecules, but only those of the type A_2B. We shall first of all discuss an A_2B molecule the equilibrium configuration of which is non-linear (see fig. 15a). An example of such a molecule is the water molecule (H_2O). Let x be the position of the B atom and m_B its mass, and let x_1 and x_2 be the positions of the A atoms, and m_A their mass. We shall choose our axes in such a way that at equilibrium

$$\begin{aligned} &x = y = z = 0; \\ &x_1 = b, \quad y_1 = c, \quad z_1 = 0; \\ &x_2 = -b, \quad y_2 = c, \quad z_2 = 0. \end{aligned} \tag{3.309}$$

The kinetic energy of the molecule is

$$T = \tfrac{1}{2}m_A(\dot{x}_1^2+\dot{y}_1^2+\dot{z}_1^2+\dot{x}_2^2+\dot{y}_2^2+\dot{z}_2^2)+\tfrac{1}{2}m_B(\dot{x}^2+\dot{y}^2+\dot{z}^2), \tag{3.310}$$

and the potential energy ($a^2 = b^2+c^2$, see fig. 15a)

$$U = \tfrac{1}{2}\alpha[(|\boldsymbol{x}_1-\boldsymbol{x}|-a)^2+(|\boldsymbol{x}_2-\boldsymbol{x}|-a)^2]+\tfrac{1}{2}\beta(|\boldsymbol{x}_1-\boldsymbol{x}_2|-2b)^2. \tag{3.311}$$

We could use as our coordinates, x, y, z, x_1-b, y_1-c, z_1, x_2+b, y_2-c, and z_2. If we denote these coordinates by primed letters: x', y', z', x_1', y_1', z_1', x_2', y_2' and z_2', we get in the small oscillations limit the following expression for the potential energy:

$$\begin{aligned} U = \frac{\alpha}{2a^2}\,[&b^2(x'-x_1')^2+c^2(y'-y_1')^2+2bc(y'-y_1')(x'-x_1')+b^2(x'-x_2')^2 \\ &+c^2(y'-y_2')^2-2bc(x'-x_2')(y'-y_2')]+\tfrac{1}{2}\beta(x_1'-x_2')^2. \end{aligned} \tag{3.312}$$

We note that the z', z_1', and z_2' coordinates are cyclic. This we might have expected, as the three-dimensional problem with nine degrees of freedom has six cyclic degrees of freedom (three rotational and three translational ones) while the two-dimensional problem with six degrees of freedom has three cyclic degrees of freedom (one rotational and two translational) in each case leaving us with three non-cyclic degrees of freedom.

It is convenient to introduce another set of coordinates by the equations

$$\begin{aligned} &q_1 = x', \quad q_2 = y', \quad q_3 = z', \\ &q_4 = \tfrac{1}{2}(x_1'+x_2'), \quad q_5 = \tfrac{1}{2}(x_1'-x_2'), \quad q_6 = \tfrac{1}{2}(y_1'-y_2'), \\ &q_7 = \tfrac{1}{2}(y_1'+y_2'), \quad q_8 = \tfrac{1}{2}(z_1'-z_2'), \quad q_9 = \tfrac{1}{2}(z_1'+z_2'). \end{aligned} \tag{3.313}$$

To see the importance of these coordinates we consider two configurations of the molecule such as depicted in fig. 15b, where

$$\begin{aligned} &\bar{x} = -x, \quad \bar{x}_1 = -x_2, \quad \bar{x}_2 = -x_1; \\ &\bar{y} = y, \quad \bar{y}_1 = y_2, \quad \bar{y}_2 = y_1; \\ &\bar{z} = z, \quad \bar{z}_1 = z_2, \quad \bar{z}_2 = z_1 \quad (\text{or } \bar{z} = -z, \quad \bar{z}_1 = -z_2, \quad \bar{z}_2 = -z_1). \end{aligned} \tag{3.314}$$

From symmetry considerations it follows immediately that the potential energy is the same in both cases. This means that the potential energy must be invariant under the following transformation

$$\begin{aligned} &q_1 \to -q_1, \quad q_2 \to q_2, \quad q_3 \to q_3 \quad (\text{or } q_3 \to -q_3), \\ &q_4 \to -q_4, \quad q_5 \to q_5, \quad q_8 \to q_8 \quad (\text{or } q_8 \to -q_8), \\ &q_6 \to -q_6, \quad q_7 \to q_7, \quad q_9 \to q_9 \quad (\text{or } q_9 \to -q_9), \end{aligned} \tag{3.315}$$

or, that the potential energy may not contain cross terms between the following three groups of variables (q_1, q_4, q_6), (q_2, q_5, q_7), and (q_3, q_8, q_9). This means that the secular equation can be factorised which enables us to find the eigenvalues and the normal modes of vibration.

Expressed in the q_i the potential energy is of the form

$$\begin{aligned} U = (\alpha/a^2)[b^2(q_1^2+q_4^2+q_5^2-2q_1q_4)+c^2(q_2^2+q_6^2+q_7^2-2q_2q_7) \\ -2bc(q_1q_6+q_2q_5-q_4q_6-q_5q_7)]+2\beta q_5^2, \end{aligned} \tag{3.316}$$

and the secular equation can be factorised into the following three ones:

$$\text{(i)} \quad \begin{vmatrix} \lambda\nu-1 & 1 & \gamma \\ 1 & \lambda\mu-1 & -\gamma \\ \gamma & -\gamma & \lambda\mu-\gamma^2 \end{vmatrix} = 0,$$

$$\text{(ii)} \quad \begin{vmatrix} \lambda\nu-\gamma^2 & \gamma & \gamma^2 \\ \gamma & \lambda\mu-1-\beta' & -\gamma \\ \gamma^2 & -\gamma & \lambda\mu-\gamma^2 \end{vmatrix} = 0, \tag{3.317}$$

$$\text{(iii)} \quad \begin{vmatrix} \lambda m_{\mathrm{B}} & 0 & 0 \\ 0 & 2\lambda m_{\mathrm{A}} & 0 \\ 0 & 0 & 2\lambda m_{\mathrm{A}} \end{vmatrix} = 0,$$

where

$$\mu = m_{\mathrm{A}}a^2/\alpha b^2, \quad \gamma = c/b, \quad \nu = m_{\mathrm{B}}\mu/2m_{\mathrm{A}}, \quad \beta' = \beta a^2/\alpha b^2. \tag{3.318}$$

In the normal vibrations corresponding to equation (3.317 i) only q_1, q_4 and q_6 are different from zero, in those corresponding to (3.317 ii) only q_2, q_5, and q_7 do not vanish, and q_3, q_8, and q_9 are non-vanishing in case (iii).

One finds that all λ's of group (iii) are zero, corresponding to the fact that the three z-coordinates are cyclic. The normal coordinates can be chosen to correspond to a translation in the z-direction and to rotations around the x- and y-axes.

Of the λ's of group (i) two are zero, corresponding to a translation along the x-axis and to a rotation around the z-axis. It is instructive to verify that, indeed, for those two motions q_2 and q_5 vanish (compare figs. 16a and b). The third λ, which is different from zero, corresponds to a normal vibration of the type depicted by fig. 16c.

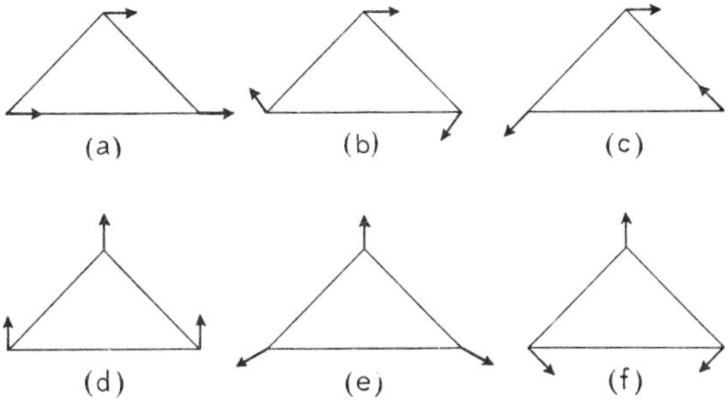

Fig. 16. The normal modes of a non-linear triatomic molecule in a plane; (a), (b), and (c) correspond to group (3.317 i); (d), (e), and (f) correspond to group (3.317 ii).

Of the λ's of group (ii) one is zero, corresponding to a translation in the y-direction (fig. 16d), while two are non-vanishing (figs. 16e and 16f). We leave it to the reader to evaluate the normal coordinates for the different cases, a tedious, but straightforward exercise.

We can now go over to linear A_2B molecules (such as CO_2). At first sight we should expect that there will be four non-vanishing eigenvalues, as there are now only two cyclic coordinates corresponding to rotations. This corresponds, of course, to the five degrees of freedom of a rigid linear structure such as a dumbbell molecule. This was found to be the case for the diatomic molecule considered at the beginning of this section. We might, moreover, also expect that we could get the behaviour of a linear molecule by letting c go to zero in the preceding analysis. It turns out, however, that instead of getting four non-vanishing eigenvalues, we get only two non-vanishing eigenvalues. In fig. 17 we have drawn the six normal modes of a linear A_2B molecule in a plane which correspond to the six modes of fig. 16. The eigenvalues corresponding to fig. 17a, 17b, 17d, and 17f are now equal to zero, while the eigenvalues corresponding to motions out of the plane remain equal to zero. The reason for this behaviour is the following one. The mode of fig. 17f corresponds to displacements such that the change in the distance between any two of the three atoms will be quadratic in the displacement of the atoms, and the first non-vanishing terms in the potential energy (3.311) will thus be quartic, and have therefore been neglected in the small amplitude approximation.

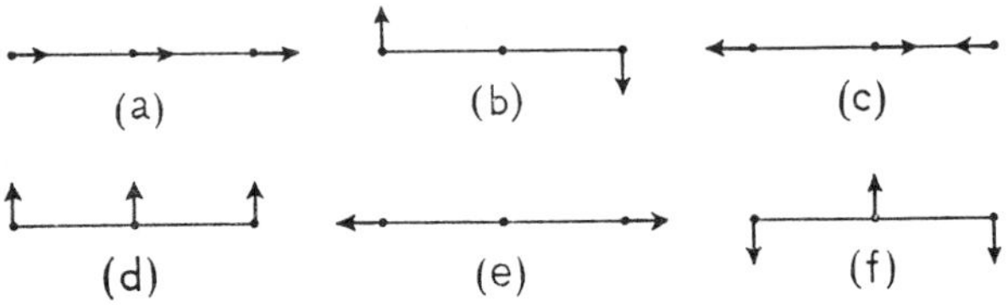

Fig. 17. The normal modes of a linear triatomic molecule in a plane.

We know, however, that bending modes, such as the one of fig. 17f occur with a non-vanishing frequency and we must thus investigate non-central forces. The simplest case to investigate is the following one, which has been used by other authors:

$$U = \tfrac{1}{2}\alpha[(x_1'-x')^2+(x_2'-x')^2] + \tfrac{1}{2}\beta[(y_1'-y')^2+(z_1'-z')^2+(y_2'-y')^2+(z_2'-z')^2]+\tfrac{1}{2}\gamma(x_1'-x_2')^2, \quad (3.319)$$

where the α, β, and γ are constants, and where the x', y', z', ... are the same coordinates as we used in (3.312). The secular equation can always be factorised in the case of linear molecules, each factor corresponding to one of the three coordinate axes. Introducing again the q_i of (3.313) and now grouping q_1, q_4, q_5 together, and also q_2, q_6, q_7, and q_3, q_8, q_9, we get three groups of eigenvalues, the first three eigenvalues are the solutions of the equation

$$\begin{vmatrix} \lambda m_B - 2\alpha & -2\alpha & 0 \\ -2\alpha & 2\lambda m_A - 2\alpha & 0 \\ 0 & 0 & 2\lambda m_A - 2\alpha - 4\gamma \end{vmatrix} = 0, \quad (3.320)$$

and the other two groups contain the same triplet of eigenvalues, corresponding to the fact that the problem is now symmetric in y and z. These three, two-fold degenerate eigenvalues are the solutions of the equation

$$\begin{vmatrix} \lambda m_B - 2\beta & 0 & -2\beta \\ 0 & 2\lambda m_A - 2\beta & 0 \\ -2\beta & 0 & 2\lambda m_A - 2\beta \end{vmatrix} = 0. \quad (3.321)$$

One finds now *six* non-vanishing eigenvalues, namely, those corresponding to figs. 17c and 17e (solution of 3.320) † and those corresponding to figs. 17b and 17f (solutions of (3.321); these eigenvalues are two-fold degenerate). The fact that we now have two non-vanishing eigenvalues too many is due to the fact that the potential (3.319) is no longer invariant under a rotation, which means that it is no longer physically meaningful. This can be remedied by considering the following potential

$$U = \tfrac{1}{2}\alpha[(x_1'-x')^2+(x_2'-x')^2]+\tfrac{1}{2}\beta[(y_1'-y')^2+(z_1'-z')^2+(y_2'-y')^2 +(z_2'-z')^2-\tfrac{1}{2}(y_1'-y_2')^2-\tfrac{1}{2}(z_1'-z_2')^2]+\tfrac{1}{2}\gamma(x_1'-x_2')^2. \quad (3.322)$$

Now, only the two-fold degenerate eigenvalue corresponding to fig. 17b vanishes, and we are, indeed, left with four non-vanishing eigenvalues. The secular equation is sufficiently simple to allow an easy solution for the normal coordinates, but we shall leave this as an exercise for the reader.

4. THE NORMAL VIBRATIONS OF A ONE-DIMENSIONAL CRYSTAL

As a last example of small vibrations around an equilibrium position we shall consider the case of a one-dimensional crystal. This is a system of N mass points which at equilibrium will be assumed to be equally spaced. We shall first of all consider the case where all the mass points have the same mass M and after that discuss the case where two kinds of mass points with masses M and m alternate. We do not have the space here to discuss all aspects of this problem — which plays an important part in the determination of the thermodynamic properties of solids — but refer to the literature for such a discussion ††.

In order to avoid end effects we assume the crystal' to be arranged on a circle so that the N-th mass point is the neighbour of the first one. This is equivalent to considering an infinite crystal on which a periodic boundary condition,

$$q_{j+N} = q_j, \quad (3.401)$$

(q_j is the displacement of the j-th mass point from its equilibrium position) is imposed.

† Since our groups of q_i are different from the groups used in (3.317), the modes corresponding to Figures 17a, 17c, and 17e, correspond to (3.320) and those of Figures 17b, 17d, and 17f to (3.321).

†† See, for instance, G. H. Wannier, Elements of Solid State Theory (Cambridge University Press, 1959).

In the case of all masses being equal, the kinetic energy of the system is given by the expression

$$T = \tfrac{1}{2}M \sum_j \dot{q}_j^2, \tag{3.402}$$

and if we assume the forces between the mass points to be harmonic and to be acting only between nearest neighbours, the potential energy is of the form

$$U = \tfrac{1}{2}\alpha \sum_j (q_j - q_{j+1})^2. \tag{3.403}$$

We see that the potential energy is quadratic in the q_j so that we can apply the theory of small vibrations developed in the first section of this chapter. One can, indeed, use known methods from the theory of determinants to determine the eigenvalues and hence the normal coordinates. It is, however, more convenient to use the fact that we should expect normal vibrations with wavelengths ranging all the way from one lattice spacing to twice the length of the crystal. We introduce therefore a set of coordinates Q_k defined by the equations

$$Q_k = \sqrt{\frac{M}{N}} \sum_j \mathrm{e}^{2\pi i kj/N} q_j. \tag{3.404}$$

If we use the relation

$$\sum_{k=1}^{N} \mathrm{e}^{2\pi i (j-m)k/N} = N\delta_{jm}, \tag{3.405}$$

we can solve (3.404) for the q_j and find

$$q_j = \frac{1}{\sqrt{MN}} \sum_k \mathrm{e}^{-2\pi i kj/N} Q_k. \tag{3.406}$$

Substituting this expression for the q_j into the expressions for T and U, and using (3.405), we get

$$T = \tfrac{1}{2} \sum_k \dot{Q}_k \dot{Q}_{-k}, \tag{3.407}$$

$$U = \tfrac{1}{2} \frac{4\alpha}{M} \sum_k Q_k Q_{-k} \sin^2 \frac{\pi k}{N}. \tag{3.408}$$

We notice that although we have not yet completely reduced T and U to the forms (3.138) and (3.139), the equations of motion are of the form (3.115); the equations of motion for Q_k follows from the Lagrangian equation for Q_{-k} and vice versa. The eigenfrequencies are given by the equation

$$\omega_k = 2(\alpha/M)^{\frac{1}{2}} \sin (\pi|k|/N). \tag{3.409}$$

We must draw attention to the fact that the Q_k are complex quantities and that thus each Q_k corresponds to two degrees of freedom. The set of the Q_k does not overdetermine the system, however, as there exists the relation

$$Q_k = Q^*_{-k}. \tag{3.410}$$

We also see from (3.404) that the Q_k satisfy an equation similar to (3.401) for the q_j, namely,

$$Q_k = Q_{k+N}, \tag{3.411}$$

and we can thus restrict k to the interval (we shall assume N to be even)

$$0 \leqq k \leqq \tfrac{1}{2}N, \tag{3.412}$$

and have $\frac{1}{2}N+1$ independent Q_k corresponding to N independent real coordinates [note that it follows from (3.410) and (3.411) that Q_0 and $Q_{\frac{1}{2}N}$ are real] for which we can choose the real and imaginary parts of the Q_k:

$$Q_k = R_k + \mathrm{i}S_k. \tag{3.413}$$

In terms of the R_k and S_k the kinetic and potential energies become

$$T = \tfrac{1}{2}\sum_k (\dot{R}_k^2 + \dot{S}_k^2), \qquad U = \tfrac{1}{2}\sum_k \omega_k^2(R_k^2 + S_k^2), \tag{3.414}$$

and the reduction to sums of squares is now complete.

To see the physical nature of the normal vibrations we use (3.406) in which we now put all Q_k, but one, equal to zero. The non-vanishing Q_k is put equal to $A \exp(\mathrm{i}\omega_k t)$, and we get

$$q_j(t) = (MN)^{-\frac{1}{2}} A \exp 2\pi\mathrm{i}[\nu_k t - (kj/N)], \qquad 2\pi\nu_k = \omega_k. \tag{3.415}$$

From (3.415) we see that the normal vibrations are travelling waves with a wave number equal to k/aN, where a is the lattice spacing.

From (3.409) it follows first of all that there is one vanishing eigenfrequency, ω_0, corresponding to the translation of the whole 'crystal'. Secondly, we see that for small k we have approximately

$$\omega_k = 2\pi k \alpha^{\frac{1}{2}}/NM^{\frac{1}{2}}, \tag{3.416}$$

corresponding to a sound wave spectrum (compare the discussion in Chapter 8). Finally, we see that the largest eigenfrequency $\omega_{\frac{1}{2}N}$ corresponds to a wavelength equal to $2a$, that is, the case where two neighbouring mass points are always moving in opposite directions.

If we had used the R_k and S_k instead of the Q_k to describe the normal modes, we would have found instead of travelling waves standing waves.

The case where the 'crystal' consists of a linear chain of alternating masses is more complicated. The kinetic and potential energies are now given by the equations

$$T = \tfrac{1}{2}M \sum_j \dot{q}_{2j}^2 + \tfrac{1}{2}m \sum_j \dot{q}_{2j+1}^2 , \tag{3.417}$$

$$U = \tfrac{1}{2}\alpha \sum_j [(q_{2j}-q_{2j+1})^2 + (q_{2j}-q_{2j-1})^2]. \tag{3.418}$$

Instead of one set of equations (3.404) or (3.406) we introduce two sets [in the following we shall throughout use complex variables and the R_k are now complex variables rather than real ones as they were in (3.413)],

$$q_{2j} = \sqrt{\frac{2}{NM}} \sum_k \mathrm{e}^{-4\pi i jk/N} Q_k , \tag{3.419}$$

$$q_{2j+1} = \sqrt{\frac{2}{Nm}} \sum_k \mathrm{e}^{-2\pi i(2j+1)k/N} R_k . \tag{3.420}$$

In terms of the Q_k and the R_k we find for T and U

$$T = \tfrac{1}{2} \sum_k (\dot{Q}_k \dot{Q}_{-k} + \dot{R}_k \dot{R}_{-k}), \tag{3.421}$$

$$U = \tfrac{1}{2}\alpha \sum_k \left[\frac{2Q_k Q_{-k}}{M} + \frac{2R_k R_{-k}}{m} - \frac{4(R_k Q_{-k} + R_{-k} Q_k)}{(mM)^{\frac{1}{2}}} \cos\frac{2\pi k}{N}\right] . \tag{3.422}$$

We see that once again the modes with different values of k are uncoupled, but the Q_k and R_k with the same k are still coupled. The eigenvalue problem corresponding to a given value of k is

$$\begin{vmatrix} \dfrac{2\alpha}{M} - \omega_k^2 & -\dfrac{2\alpha}{\sqrt{mM}} \cos\dfrac{2\pi k}{N} \\ -\dfrac{2\alpha}{\sqrt{mM}} \cos\dfrac{2\pi k}{N} & \dfrac{2\alpha}{m} - \omega_k^2 \end{vmatrix} = 0, \tag{3.423}$$

from which we get for the ω_k^2

$$\omega_k^2 = \alpha \left[\frac{1}{M} + \frac{1}{m} \pm \left\{\left(\frac{1}{M} + \frac{1}{m}\right)^2 - \frac{4}{mM} \sin^2 \frac{2\pi k}{N}\right\}^{\frac{1}{2}}\right] . \tag{3.424}$$

For a given value of ω_k the ratio of Q_k/R_k follows in the usual way and we have

$$\frac{Q_k}{R_k} = \sqrt{\frac{M}{m}}\frac{\alpha\cos(2\pi k/N)}{\alpha-\frac{1}{2}M\omega_k^2} = \sqrt{\frac{M}{m}}\frac{\alpha-\frac{1}{2}m\omega_k^2}{\alpha\cos(2\pi k/N)}. \tag{3.425}$$

From (3.424) and (3.425) we can find the physical nature of the normal modes. If we take the lower sign in (3.424), ω_k^2 is always less than $2\alpha/m$ and less than $2\alpha/M$ so that Q_k and R_k have the same sign: two neighbouring masses will move in phase [see (3.419) and (3.420)]. As $k \to 0$, ω_k is proportional to k and we are dealing with sound waves. The branch of the eigenfrequency spectrum corresponding to the lower sign is for this reason called the *acoustical branch.*

If we take the upper branch, ω_k^2 is always larger than $2\alpha/m$ and than $2\alpha/M$, and two neighbouring masses will move in opposite directions. If the two masses carry different charges as will, for instance, be the case in an alkali-halide crystal, the normal vibrations will correspond to a vibrating dipole-moment. The branch corresponding to the upper sign is therefore called the *optical branch.* In the alkali-halides this branch corresponds to the so-called 'Reststrahlen' band in the infra-red.

We leave a discussion of other features of the normal vibrations of a crystal to the reader.

5. OSCILLATIONS AROUND AN EQUILIBRIUM MOTION

There are many problems where one is interested in discussing the stability, not of an equilibrium position, but of an equilibrium motion. These problems have in recent years come to the fore in the discussion of particle orbits in high energy accelerators.

We shall consider here the simple two-dimensional case † of a particle moving in a circular orbit under the influence of a central field potential $U(r)$ given by the equation

$$U = -Ar^{-\alpha}, \tag{3.501}$$

where A and α are constants and where r is the distance from the force centre. We may ask whether such a motion in a circle of radius r_0 is a stable one. The Lagrangian is of the form

$$L = \tfrac{1}{2}m(\dot{r}^2 + r^2\dot{\theta}^2) + Ar^{-\alpha}, \tag{3.502}$$

† H. Goldstein, Classical Mechanics (Addison-Wesley, Cambridge, Mass., 1950) p. 346.

where θ is the polar angle, and the equations of motion are

$$\begin{aligned} m\ddot{r} - mr\dot{\theta}^2 + \alpha A r^{-\alpha-1} &= 0 \\ mr^2\ddot{\theta} + 2mr\dot{r}\dot{\theta} &= 0. \end{aligned} \tag{3.503}$$

If we are dealing with a circular motion $\dot{r} = 0$ so that we find from the second equation that $\dot{\theta} = \text{constant} = \omega_0$, and from the first equation it follows that

$$mr_0\omega_0^2 = \alpha A r_0^{-\alpha-1}. \tag{3.504}$$

We now consider oscillations around the equilibrium motion which is characterised by r_0 and ω_0, which are connected through (3.504). In order to do that we write

$$r = r_0 + q_1, \qquad \dot{\theta} = \omega_0 + q_2. \tag{3.505}$$

Substituting (3.505) into (3.503) and neglecting second order terms in the q_i we get [the zeroth order terms cancel by virtue of (3.504)],

$$\begin{aligned} m\ddot{q}_1 - m\omega_0^2 q_1 - 2mr_0\omega_0 q_2 - \alpha(\alpha+1)Ar_0^{-\alpha-2}q_1 &= 0, \\ mr_0^2\dot{q}_2 + 2mr_0\omega_0\dot{q}_1 &= 0. \end{aligned} \tag{3.506}$$

If we now assume that q_1 and q_2 are periodic in time with the same frequency:

$$q_i = A_1 e^{i\omega t}, \qquad q_2 = A_2 e^{i\omega t}, \tag{3.507}$$

which is analogous to (3.116), we get the following secular equation for ω,

$$\begin{vmatrix} m(-\omega^2 - \omega_0^2) - \alpha(\alpha+1)Ar_0^{-\alpha-2} & -2mr_0\omega_0 \\ 2mr_0\omega_0\omega & mr_0^2\omega \end{vmatrix} = 0. \tag{3.508}$$

We find one zero root which corresponds to a slightly displaced equilibrium circular orbit, where the $(r_0 + A_1)$ and $(\omega_0 + A_2)$ satisfy the same equation as r_0 and ω_0 did before, that is, (3.504). For the other roots we find the equation

$$\omega^2 = (2-\alpha)\omega_0^2. \tag{3.509}$$

We see therefore that, provided α is less than 2, the orbit will be a stable one.

We may draw attention here to the fact that the secular equation (3.508) is more complicated than (3.121) in so far as ω occurs not only as ω^2, but also as ω. This is a consequence of the fact that we are no longer discussing equilibrium positions, but equilibrium motions.

CHAPTER 4

DYNAMICS OF RIGID BODIES

After defining the Euler angles, the two equations of motion for rigid bodies, one describing its translational and the other describing its rotational motion, are derived. An expression is derived for the kinetic energy of a rigid body in terms of its moments of inertia and the rotational velocities referred to its principal axes. The Euler equations are derived and applied to the case of rigid bodies on which no external forces act and to the case of the heavy symmetric top. The precession and nutation of the earth's axis due to the solar and lunar gravitational forces is discussed. In the last section we discuss the Coriolis force and its influence on a freely falling mass and on a spherical pendulum (Foucault's pendulum).

1. INTRODUCTION

Up to now we have considered systems of point particles which were in general of such a nature that each of the particles could move to some extent independently. In the present chapter we shall be concerned with systems of point particles which are not collinear and which are such that the distance between any two particles is rigidly fixed. Such systems are called *rigid bodies.*

A rigid body has six degrees of freedom. This can be seen in two ways. Firstly, we would expect three translational and three rotational degrees of freedom. This intuitive approach tallies, indeed, with a more rigorous discussion. Consider to begin with three non-collinear particles. The position of each of these particles is determined by three coordinates, but there are three constraints of the type (2.108) which leaves us with six degrees of freedom (compare the discussion in § 3.3 of the non-linear triatomic molecule). The position of each further particle in the system is again determined by three coordinates, but the distances of this particle from the first three particles are fixed, and there are therefore no more degrees of freedom.

We shall use as the six generalised coordinates to determine the configuration of the rigid body the three centre of mass coordinates X, Y, and Z, and the three angles θ, φ, and ψ which will characterise the orientation of a set of mutually perpendicular axes in space. It is clear that the first three coordinates correspond to the translational degrees of freedom, while the second triplet corresponds to the rotational degrees of freedom. To define the angles we choose a set of coordinate axes, X, Y, Z, fixed in the body, while x, y, z is a set of axes fixed in space (see fig. 18). The angle θ is defined

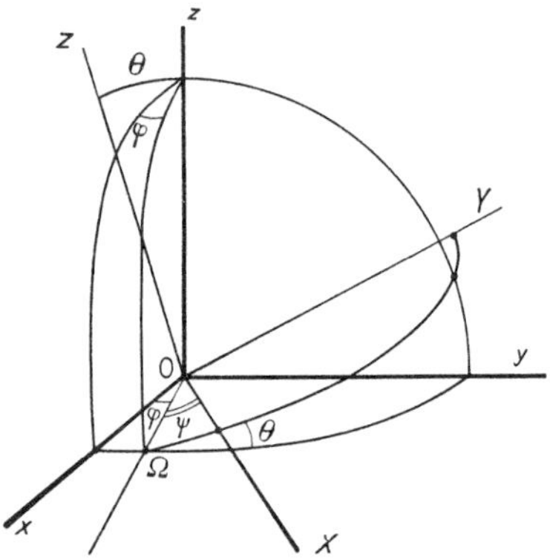

Fig. 18. The Euler angles. The XYZ-system is fixed in the body; the xyz-system in space.

as the angle between the z-axis and the Z-axis. The angle φ is the angle between the x-axis and the line of nodes OΩ, which is the line of intersection of the Oxy- and the OXY-planes. Finally, the angle ψ is the angle between the line of nodes and the X-axis. The angles defined in this way are the so-called Euler angles. We must warn the reader that unfortunately there is no unanimity in the literature on the definition or notation of the Euler angles. The definition used here is the one most generally used, but it is advisable to check the definition of the Euler angles before comparing expressions in two different treatises on the subject.

For future reference it is convenient to have expressions for the cosines of the angles between one of the xyz-axes and one of the XYZ-axes. These expressions are easily obtained from fig. 18, if one uses the cosine and sine laws of spherical trigonometry:

$$\cos a = \cos b \cos c + \sin b \sin c \cos A, \tag{4.101}$$

$$\frac{\sin a}{\sin A} = \frac{\sin b}{\sin B} = \frac{\sin c}{\sin C}, \tag{4.102}$$

where a, b, and c are the lengths of the arcs of the spherical triangle ABC, while A, B, and C are the angles between the arcs (see fig. 19).

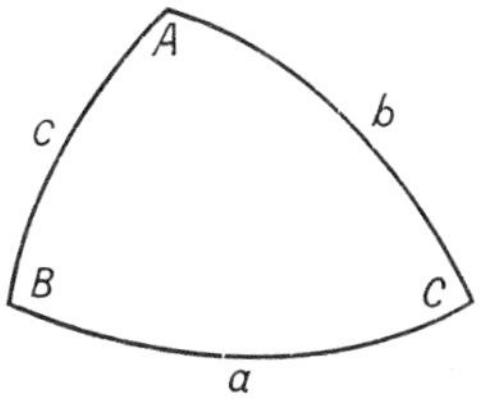

Fig. 19. A spherical triangle.

One finds then the results of table 1. Each entry gives the value of the cosine between the axes mentioned at the top of the corresponding column and to the left of the corresponding row. It may be remarked here that one can also obtain the results of Table 1 — be it less straightforwardly — by observing that the transformation from XYZ-axes to xyz-axes can be split into three successive rotations: through an angle ψ around the z-axis, through an angle θ around the x-axis, and through an angle φ around the z-axis.

TABLE 1

Values of the cosines of the angles between the axes fixed in the body and the axes fixed in space, expressed in the Eulerian angles

	X	Y	Z
x	$\cos\psi\cos\varphi - \cos\theta\sin\psi\sin\varphi$	$-\sin\psi\cos\varphi - \cos\theta\cos\psi\sin\varphi$	$\sin\theta\sin\varphi$
y	$\cos\psi\sin\varphi + \cos\theta\sin\psi\cos\varphi$	$-\sin\psi\sin\varphi + \cos\theta\cos\psi\cos\varphi$	$-\sin\theta\cos\varphi$
z	$\sin\psi\sin\theta$	$\cos\psi\sin\theta$	$\cos\theta$

In the following it will be convenient to express the motion of the rigid body in terms of the angular velocities around the X, Y, and Z-axes instead of in terms of $\dot{\theta}$, $\dot{\varphi}$, and $\dot{\psi}$. However, as the $\dot{\theta}$, $\dot{\varphi}$, and $\dot{\psi}$ will enter into the Lagrangian equations of motion, it will be necessary to find the relations between the two sets of angular velocities. If p, q, and r (or ω_1, ω_2, ω_3) are the angular velocities around the X, Y, and Z-axes we find

$$\begin{aligned} \omega_1 &= p = \dot{\theta}\cos\psi + \dot{\varphi}\sin\theta\sin\psi, \\ \omega_2 &= q = -\dot{\theta}\sin\psi + \dot{\varphi}\sin\theta\cos\psi, \\ \omega_3 &= r = \dot{\varphi}\cos\theta + \dot{\psi}. \end{aligned} \tag{4.103}$$

Equations (4.103) can be obtained, if one bears in mind that $\dot{\theta}$, $\dot{\varphi}$, and $\dot{\psi}$ are the angular velocities around the line of nodes, OΩ, around the z-axis, and around the Z-axis, respectively.

We shall now find the equations of motion for a rigid body in a frame of reference which is fixed in space, that is, in an inertial system. We shall afterwards want them also in a frame of reference which is fixed in the body. To derive these equations of motion we shall use d'Alembert's principle (2.226),

$$\sum_i (m_i\ddot{\boldsymbol{x}}_i - \boldsymbol{F}_i \cdot \delta\boldsymbol{x}_i) = 0, \tag{4.104}$$

where the $\boldsymbol{F}_i$ are the external forces acting on the particles, and where the $\delta\boldsymbol{x}_i$ are such that the kinematic relations are not violated. This last requirement is a very restrictive one and means in fact that only two types of $\delta\boldsymbol{x}_i$ are possible:

$$\text{(i) } \delta\boldsymbol{x}_i = \boldsymbol{\varepsilon}, \quad \text{and} \quad \text{(ii) } \delta\boldsymbol{x}_i = [\boldsymbol{\varepsilon} \wedge \boldsymbol{x}_i], \tag{4.105}$$

corresponding respectively to translations and to rotations of the body.

If we combine (4.104) and (4.105 (i)) we get

$$(\boldsymbol{\varepsilon} \cdot \sum_i m_i\ddot{\boldsymbol{x}}_i) = (\boldsymbol{\varepsilon} \cdot \sum_i \boldsymbol{F}_i), \tag{4.106}$$

or, since $\boldsymbol{\varepsilon}$ is an arbitrary vector,

$$M\ddot{\boldsymbol{X}}_{\text{com}} = \dot{\boldsymbol{P}} = \boldsymbol{F}_{\text{res}}, \tag{4.107}$$

where M is the total mass of the system,

$$M = \sum_i m_i, \tag{4.108}$$

$\boldsymbol{X}_{\text{com}}$ the vector whose components are the centre of mass coordinate,

$$\boldsymbol{X}_{\text{com}} = \sum_i m_i\boldsymbol{x}_i/M, \tag{4.109}$$

and $\boldsymbol{F}_{\text{res}}$ the resultant of the forces acting on the body,

$$\boldsymbol{F}_{\text{res}} = \sum_i \boldsymbol{F}_i. \tag{4.110}$$

On the other hand, if we combine (4.104) and (4.105 (ii)) we get

$$(\boldsymbol{\varepsilon} \cdot \sum_i [\boldsymbol{x}_i \wedge m_i \ddot{\boldsymbol{x}}_i]) = (\boldsymbol{\varepsilon} \cdot \sum_i [\boldsymbol{x}_i \wedge \boldsymbol{F}_i], \tag{4.111}$$

or, since $\boldsymbol{\varepsilon}$ once again is arbitrary,

$$\dot{\boldsymbol{J}} = \boldsymbol{\mathcal{M}}, \tag{4.112}$$

where $\boldsymbol{J}$ is the total angular momentum of the body,

$$\boldsymbol{J} = \sum_i [\boldsymbol{x}_i \wedge m_i \dot{\boldsymbol{x}}_i], \tag{4.113}$$

and $\boldsymbol{\mathcal{M}}$ the total moment of the forces acting on the body,

$$\boldsymbol{\mathcal{M}} = \sum_i [\boldsymbol{x}_i \wedge \boldsymbol{F}_i]. \tag{4.114}$$

Equations (4.107) and (4.112) are the basic equations for the motion of a rigid body. The first one expresses the fact that the centre of mass of a rigid body moves as if all the mass were concentrated in that point and all forces were acting there. The second equation gives an expression for the rate of change of the angular momentum which equals the total moment of all forces acting on the body — both the angular momentum and the total moment being evaluated with respect to the same point of reference, for which we chose the origin in (4.113) and (4.114).

Let us consider for a moment a pure rotation of the rigid body. The instantaneous rotation may be characterised by the vector $\boldsymbol{\omega}$, whose components are given by (4.103). The velocities of the constituent particles are given by the equation [compare (4.105 (ii))]

$$\dot{\boldsymbol{x}}_i = [\boldsymbol{\omega} \wedge \boldsymbol{x}_i]. \tag{4.115}$$

From (4.113) and (4.115) it follows that $[r_i^2 = (\boldsymbol{x}_i \cdot \boldsymbol{x}_i)]$

$$\boldsymbol{J} = \sum_i m_i[\boldsymbol{x}_i \wedge [\boldsymbol{\omega} \wedge \boldsymbol{x}_i]] = (\sum_i m_i r_i^2)\boldsymbol{\omega} - \sum_i m_i \boldsymbol{x}_i(\boldsymbol{x}_i \cdot \boldsymbol{\omega}). \tag{4.116}$$

We can write this equation as follows

$$J_k = \sum_{l=x,y,z} D_{kl}\omega_l, \qquad k = x, y, z, \tag{4.117}$$

where

$$D_{kl} = \sum_i m_i(r_i^2 \delta_{kl} - x_{ik} x_{il}), \qquad k, l = x, y, \text{ or } z. \tag{4.118}$$

This tensor of the second rank, D_{kl}, is the inertial tensor. Its diagonal elements are called *moments of inertia* and its off-diagonal elements *moments of deviation*. The off-diagonal elements with an opposite sign are also called *products of inertia*. It is instructive to go over to the case of a continuous mass distribution. This means that we must change a sum over m_i to an integral over the mass density ρ. We then get for the components of the inertial tensor,

$$\begin{aligned} D_{xx} &= \int \rho(y^2+z^2)\mathrm{d}^3\boldsymbol{x}, & D_{xy} &= -\int \rho xy\,\mathrm{d}^3\boldsymbol{x} = D_{yx}, \\ D_{yy} &= \int \rho(z^2+x^2)\mathrm{d}^3\boldsymbol{x}, & D_{yz} &= -\int \rho yz\,\mathrm{d}^3\boldsymbol{x} = D_{zy}, \\ D_{zz} &= \int \rho(x^2+y^2)\mathrm{d}^3\boldsymbol{x}, & D_{zx} &= -\int \rho zx\,\mathrm{d}^3\boldsymbol{x} = D_{xz}, \end{aligned} \tag{4.119}$$

where $\mathrm{d}^3\boldsymbol{x}$ is the volume element $\mathrm{d}x\mathrm{d}y\mathrm{d}z$.

The reason for the term 'moment of deviation' is that if the off-diagonal elements are non-vanishing there will be a 'force' tending to change the direction of $\boldsymbol{J}$ [compare (4.205)] — unless $\boldsymbol{J}$ is along one of the three principal axes (vide infra).

It is well known from the theory of second rank tensors that it is possible to find such an orientation of the axes that the inertial tensor becomes a diagonal tensor. A transformation to such axes is called a principal axis transformation and the tensor is said to be brought on its principal axes. We must emphasise that as in general the orientation of the rigid body changes with time, so does the orientation of the principal axes in space. Unless it is explicitly stated otherwise we shall assume the XYZ-axes which are fixed in the body to be along the three principal axes. The fact that these axes will remain along the principal axes indicates the possible advantages of describing the motion of the body in the XYZ-system rather than in the xyz-system which is fixed in space.

If (4.117) refers to a system of axes which are the principal axes, it reduces to

$$J_k = D_k \omega_k, \tag{4.120}$$

or

$$J_x = Ap, \qquad J_y = Bq, \qquad J_z = Cr, \tag{4.121}$$

where we have denoted by A, B, and C the three *principal moments of inertia*.

Using (4.115) we can obtain an expression for the kinetic energy of the rotational motion, as follows,

$$T = \sum_i \tfrac{1}{2} m_i \dot{x}_i^2 = \sum_i \tfrac{1}{2} m_i ([\omega \wedge x_i] \cdot [\omega \wedge x_i]) = \tfrac{1}{2} \sum_{k,l} D_{kl} \omega_k \omega_l, \qquad (4.122)$$

or, if the xyz-axes are along the principal axes,

$$T = \tfrac{1}{2} A p^2 + \tfrac{1}{2} B q^2 + \tfrac{1}{2} C r^2. \qquad (4.123)$$

If the body is moving with a linear velocity $\boldsymbol{v}$ as well as with a rotational velocity $\boldsymbol{\omega}$, we have instead of (4.115)

$$\dot{x}_i = v + [\omega \wedge x_i], \qquad (4.124)$$

and (4.122) is replaced by

$$T = \tfrac{1}{2} M v^2 + \tfrac{1}{2} \sum_{k,l} D_{kl} \omega_k \omega_l + (v \cdot [\omega \wedge M X_{\text{com}}]), \qquad (4.125)$$

where M and X_{com} are given by (4.108) and (4.109). If we choose our origin so as to be at the centre of mass, the last term on the right hand side of (4.125) vanishes and the kinetic energy becomes the sum of a translational and a rotational term:

$$T = \tfrac{1}{2} M v^2 + \tfrac{1}{2} \sum_{k,l} D_{kl} \omega_k \omega_l. \qquad (4.126)$$

2. THE EULER EQUATIONS

Equation (4.112) is the equation of motion for the rotation of a rigid body under the influence of external forces. This equation refers to a system of reference fixed in space. As we mentioned a moment ago, it will be advantageous to use the corresponding equation of motion which refers to a system of reference fixed in the body — preferably along the principal axes. To do this we must first of all find the relation between the rates of change of an arbitrary vector $\boldsymbol{a}$ in the xyz-system, fixed in space, and in the XYZ-system, fixed in the rigid body. If the second rate of change were zero, we would be dealing with a vector fixed in the body, and we should have

$$\left(\frac{\mathrm{d}a}{\mathrm{d}t}\right)_{xyz} = [\omega \wedge a]. \qquad (4.201)$$

This equation can be obtained by observing that when the body is rotated over an angle $|\varepsilon|$ around an axis parallel to $\boldsymbol{\varepsilon}$, a point with coordinates a_x, a_y, a_z will be moved to a point with coordinates $a_x + \delta a_x$, $a_y + \delta a_y$,

$a_z + \delta a_z$ in such a way that [compare (4.105 (ii))]

$$\delta \boldsymbol{a} = [\boldsymbol{\varepsilon} \wedge \boldsymbol{a}]. \tag{4.202}$$

If the vector $\boldsymbol{a}$ is itself moving with respect to the XYZ-system, we have

$$\left(\frac{\mathrm{d}\boldsymbol{a}}{\mathrm{d}t}\right)_{xyz} = \left(\frac{\mathrm{d}\boldsymbol{a}}{\mathrm{d}t}\right)_{XYZ} + [\omega \wedge \boldsymbol{a}], \tag{4.203}$$

which equation follows in the same way as (4.202) if we observe that (4.202) must be replaced by

$$\delta \boldsymbol{a} = \boldsymbol{\varepsilon}' + [\boldsymbol{\varepsilon} \wedge \boldsymbol{a}], \tag{4.204}$$

if apart from the rotation, leading to (4.201), the point $\boldsymbol{a}$ also is subject to a translation over a distance $\boldsymbol{\varepsilon}'$.

Applying (4.203) to (4.112) with $\boldsymbol{a} = \boldsymbol{J}$, and dropping the index 'XYZ' in the final result, we get for the rotational equations of motion in the XYZ-system the equations

$$\frac{\mathrm{d}\boldsymbol{J}}{\mathrm{d}t} + [\omega \wedge \boldsymbol{J}] = \mathcal{M}, \tag{4.205}$$

or, in components and using (4.121),

$$\begin{aligned} A\dot{p} + (C-B)qr &= \mathcal{M}_1, \\ B\dot{q} + (A-C)pr &= \mathcal{M}_2, \\ C\dot{r} + (B-A)pq &= \mathcal{M}_3. \end{aligned} \tag{4.206}$$

These equations are the so-called *Euler equations*.

Let us first of all consider the case $\mathcal{M} = 0$, that is, the case where there are no net torques. We emphasise once again that we are not considering the translational motion which is described by (4.107). In the case $\mathcal{M} = 0$, we have instead of (4.206)

$$\begin{aligned} A\dot{p} + (C-B)qr &= 0, \\ B\dot{q} + (A-C)pr &= 0, \\ C\dot{r} + (B-A)qp &= 0, \end{aligned} \tag{4.207}$$

where, of course, the dots refer to rates of change in the XYZ-system. We see here the advantage of using equations of motion in the XYZ-system: having once chosen our axes along the principal axes, we are sure that they will stay along the principal axes. This would not be the case, if we were working in the xyz-system.

We can immediately find two integrals of motion. Multiplying the three equations (4.207) by p, q, and r, respectively, and adding them we find

$$\tfrac{1}{2}Ap^2+\tfrac{1}{2}Bq^2+\tfrac{1}{2}Cr^2 = \text{constant}, \tag{4.208}$$

which is the energy equation [compare (4.123)].

Multiplying the equations (4.207) by Ap, Bq, and Cr, respectively, and adding them we find

$$A^2p^2+B^2q^2+C^2r^2 = \text{constant}, \tag{4.209}$$

which shews that the total angular momentum is a constant of motion. Using the two integrals of motion (4.208) and (4.209) one can express the general solution of equations (4.207) in terms of elliptic integrals. We shall not do this, but instead consider a few special cases.

First of all we note that we can find a solution of equations (4.207) of the form

$$p = \text{constant} = p_0, \qquad q = 0, \qquad r = 0, \tag{4.210a}$$

or of the form

$$q = \text{constant}, \qquad p = 0, \qquad r = 0, \tag{4.210b}$$

or

$$r = \text{constant}, \qquad p = 0, \qquad q = 0. \tag{4.210c}$$

These solutions correspond to rotations with constant angular velocity around the three principal axes. One can shew that (4.210) give the only cases where (4.207) with three unequal principal moments of inertia have a solution

$$\boldsymbol{\omega} = \text{constant},$$

or,

$$p = p_0, \qquad q = q_0, \qquad r = r_0. \tag{4.211}$$

To shew this, one substitutes (4.211) into (4.207) and finds that these equations can only be solved, if at least two out of p_0, q_0, and r_0 vanish. We may in this connexion also refer to our discussion of the term 'moment of deviation' in the preceding section.

It is interesting to study the stability of the rotations (4.210). We can do this by using the same method as was used in the discussion of small vibrations in the preceding chapter, that is, we put

$$p = p_0+\pi, \qquad q = \xi, \qquad r = \rho, \tag{4.212}$$

substitute these expressions into (4.207), and neglect terms quadratic in π, ξ, and ρ. The result is

$$\begin{aligned} A\dot{\pi} &= 0, \\ B\dot{\xi}+(A-C)p_0\rho &= 0, \\ C\dot{\rho}+(B-A)p_0\xi &= 0. \end{aligned} \tag{4.213}$$

The first equation leads to a zero-frequency small oscillation and the solution π = constant, $\xi = 0$, $\rho = 0$, that is, a solution which is again of the form (4.210a), but now with a slightly changed angular velocity around the X-axis. The last two equations of the set (4.213) lead to a small-vibrations frequency ω which is a solution of the equation

$$BC\omega^2 = (A-C)(A-B)p_0^2, \tag{4.214}$$

from which we see that the solution (4.210a) is stable provided A is either the largest or the smallest of the three principal moments of inertia. The rotations around the two principal axes corresponding to the largest and the smallest principal moments of inertia are stable equilibrium rotations, while the rotation around the third principal axis, corresponding to the intermediate principal moment of inertia, is an unstable equilibrium motion.

The second special case to be considered is the one where $A = B \neq C$. The last equation (4.207) then becomes

$$C\dot{r} = 0, \quad \text{or,} \quad r = \text{constant} = r_0. \tag{4.215}$$

Putting $A = B$ in the first two equations we have

$$\begin{aligned} A\dot{p}+(C-A)qr_0 &= 0, \\ A\dot{q}+(A-C)pr_0 &= 0, \end{aligned} \tag{4.216}$$

with the solution

$$\begin{aligned} p &= p_0 \cos[(C-A)r_0t/A], \\ q &= q_0 \sin[(C-A)r_0t/A]. \end{aligned} \tag{4.217}$$

This solution corresponds to a precession of the vector $\boldsymbol{\omega}$ around the Z-axis. This can also be seen by writing the equations of motion in the form

$$\dot{\boldsymbol{\omega}} = -[\boldsymbol{\Omega} \wedge \boldsymbol{\omega}], \tag{4.218}$$

where $\boldsymbol{\Omega}$ is a vector with components 0, 0, and $(C-A)r_0/A$.

The earth is a good example of a body with $A = B$. In that case $(C-A)/A$ is about 1/300, and r_0 is 1 day^{-1}, so that the precessional period is of the order of one year.

We shall now consider a few cases where there are torques. The first case is that of the heavy symmetric top. This is a body with $A = B$ and with one point on the axis of symmetry fixed. If l be the distance between the fixed point and the centre of mass, and m the mass of the top, its potential energy U is given by the equation

$$U = W \cos \theta, \qquad W = mgl, \tag{4.219}$$

where g is the gravitational acceleration which for this problem we shall consider to be a constant, and where we have chosen the z-axis along the vertical so that θ is one of the Euler angles.

The kinetic energy is given by the equation

$$T = \tfrac{1}{2}A(p^2+q^2)+\tfrac{1}{2}Cr^2, \tag{4.220}$$

or, if we use (4.103),

$$T = \tfrac{1}{2}A(\dot{\theta}^2+\dot{\varphi}^2 \sin^2 \theta)+\tfrac{1}{2}C(\dot{\varphi} \cos \theta+\dot{\psi})^2. \tag{4.221}$$

The Lagrangian of our problem is now

$$L = \tfrac{1}{2}A(\dot{\theta}^2+\dot{\varphi}^2 \sin^2 \theta)+\tfrac{1}{2}C(\dot{\varphi} \cos \theta+\dot{\psi})^2 - W \cos \theta, \tag{4.222}$$

and we see that there are two cyclic coordinates, φ and ψ, so that

$$\begin{aligned} p_\psi &= \text{constant} = C(\dot{\varphi} \cos \theta+\dot{\psi}), \\ p_\varphi &= \text{constant} = A\dot{\varphi} \sin^2 \theta+C \cos \theta(\dot{\varphi} \cos \theta+\dot{\psi}). \end{aligned} \tag{4.223}$$

The first of these equations is equivalent to: Cr = constant, which would have followed from the last Euler equation, since $A = B$ and $\mathcal{M}_3 = 0$.

There are three degrees of freedom so that we should expect six integration constants. We have found two of them, p_ψ and p_φ, and a third one will be the energy, E,

$$E = \tfrac{1}{2}A(\dot{\theta}^2+\dot{\varphi}^2 \sin^2 \theta)+\tfrac{1}{2}C(\dot{\varphi} \cos \theta+\dot{\psi})^2+W \cos \theta. \tag{4.224}$$

We can use (4.223) to eliminate $\dot{\varphi}$ and $\dot{\psi}$,

$$E = \tfrac{1}{2}A\dot{\theta}^2+\frac{(p_\varphi - p_\psi \cos \theta)^2}{2A \sin^2 \theta}+\frac{p_\psi^2}{2C}+W \cos \theta. \tag{4.225}$$

This equation can be integrated,

$$t = \int_{z(0)}^{z(t)} [f(z)]^{-\frac{1}{2}} dz, \tag{4.226}$$

where

$$f(z) = (1-z^2)(\alpha-az)-(\beta-bz)^2$$

$$CA\alpha = 2CE-p_{\psi}^2, \quad A\beta = p_{\varphi}, \quad Aa = 2W, \quad Ab = p_{\psi}, \tag{4.227}$$

and

$$z = \cos\theta. \tag{4.228}$$

In principle equation (4.226) gives us θ as a function of time. From (4.223) we can then find φ and ψ as functions of time. All three angles can be expressed in terms of elliptic integrals. There is not much point in discussing the complete solutions in detail, but the physical nature of the solution can be

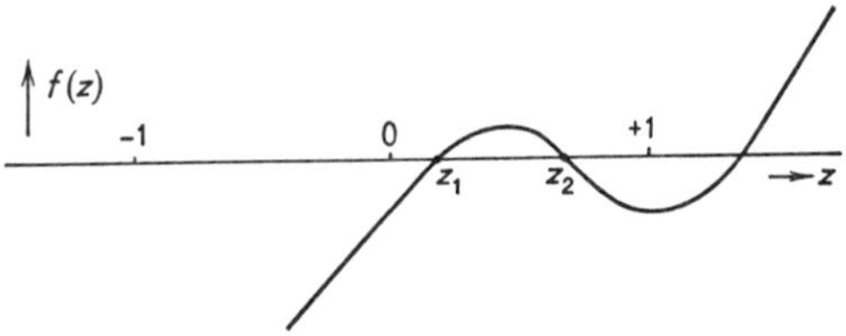

Fig. 20. The function $f(z)$ of equation (4.227).

seen from (4.226). The function $f(z)$ is a cubic polynomial in z, and its general behaviour is illustrated in fig. 20. From (4.228) it follows that z must lie between -1 and $+1$. At those points $f(z)$ is non-positive [see

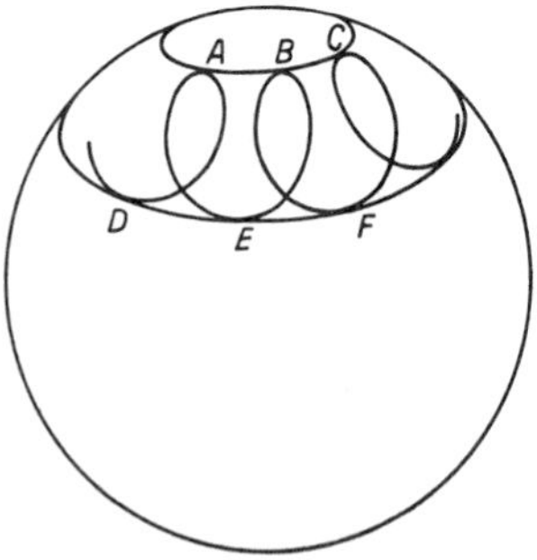

Fig. 21. Nutation and precession of a heavy symmetric top.

(4.227)]. From (4.226) it follows that $f(z)$ cannot be negative for a physical situation. This means that z will lie between z_1 and z_2, or θ between the corresponding values θ_1 and θ_2: the axis of the top will shew *nutation*.

Physically the range of possible z-values will often be much more restricted than we have discussed so far. If, for instance, we are dealing with a real top on a table, θ will have to be less than $\frac{1}{2}\pi$, and z will have to be positive.

In general $\dot{\varphi}$ will be different from zero, corresponding to a *precession*. In fig. 21 we have illustrated a possible motion. The intersection of the axis of the top with a sphere whose centre is at the point where the top is supported will describe a path such as the one depicted in fig. 21. The circle ABC corresponds to $\theta = \theta_2$ and the circle DEF to $\theta = \theta_1$.

Nutation and precession are also shown by the earth's axis. Here the torque exerted on the earth's axis is due to the gravitational forces of the sun and the moon on the earth. We shall, to begin with, only consider the influence of the sun. To solve this problem it is convenient to use equations of motion which are intermediate between the Euler equations (4.207) and (4.112). Instead of using either a system of coordinate axes fixed in space, (4.112), or one fixed in the earth, (4.207), we shall use the following system of axes (see fig. 22): we take the ζ-axis along the earth's axis, the ξ-axis along the

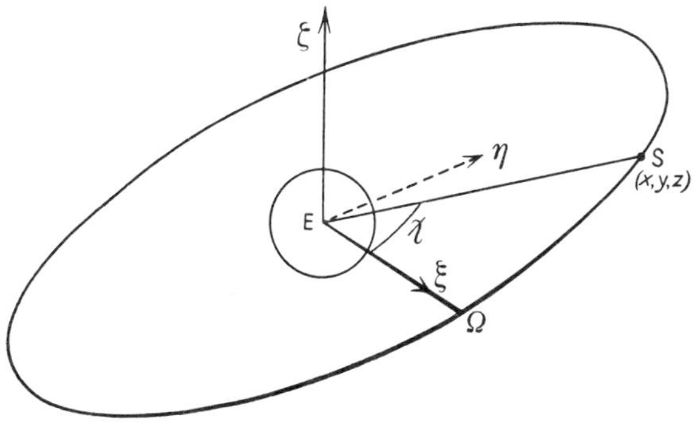

Fig. 22. The orbit of the sun around the earth. E is the centre of the earth; S is the position of the sun in its orbit around the earth; EΩ is the line of nodes.

line of nodes, that is, the intersection of the earth's equatorial plane and the plane of the earth's orbit around the sun (this plane, the ecliptic plane, is, of course, also the plane of the sun's orbit around the earth, and for our present discussion it is simpler to look at it from this latter point of view), and the η-axis in such a way that the ξ-, η-, and ζ-axes form a right handed set of orthogonal axes.

The equations of motion are now

$$\dot{\boldsymbol{J}}+[\boldsymbol{\omega}_0 \wedge \boldsymbol{J}] = \mathcal{M}, \tag{4.229}$$

where $\boldsymbol{\omega}_0$ is the rotational velocity vector of the ξ-, η-, and ζ-axes (its components in the $\xi\eta\zeta$-system will be denoted by ω_ξ, ω_η, and ω_ζ). As the ζ-axis is along the axis of symmetry of the earth, (4.120) holds and we have

$$J_\xi = Ap, \qquad J_\eta = Aq, \qquad J_\zeta = Cr, \tag{4.230}$$

where we have put $B = A$, and where p, q, and r are the rotational velocity components of the earth.

The components of the torque $\mathcal{M}$ can be found from the equation

$$\mathcal{M} = [\boldsymbol{X} \wedge \boldsymbol{F}], \tag{4.231}$$

where $\boldsymbol{X}$ is the position of the sun in the $\xi\eta\zeta$-system (components X, Y, Z) and where $\boldsymbol{F}$ is the force exerted by the sun on the earth. To find $\boldsymbol{F}$ we note that we can write it as follows

$$\boldsymbol{F} = \nabla_{\boldsymbol{X}} U(\boldsymbol{X}), \tag{4.232}$$

where U is the gravitational potential due to the earth and $\nabla_{\boldsymbol{X}}$ is the gradient vector with components $\partial/\partial X$, $\partial/\partial Y$, and $\partial/\partial Z$. In writing down (4.232) we have used Newton's third law.

The potential U is given by the equation

$$U(\boldsymbol{X}) = -GM_\odot \int \frac{\rho(\boldsymbol{x})\mathrm{d}^3\boldsymbol{x}}{|\boldsymbol{x}-\boldsymbol{X}|}, \tag{4.233}$$

where $\rho(\boldsymbol{x})$ is the density at the point $\boldsymbol{x}$ somewhere in the earth, G the gravitational constant, and $M_\odot$ the solar mass. Expanding $|\boldsymbol{x}-\boldsymbol{X}|^{-1}$ in powers of $|\boldsymbol{x}|/|\boldsymbol{X}|$ we get up to terms of order R^{-3} ($R = |\boldsymbol{X}|$):

$$U = -\frac{GMM_\odot}{R} + \frac{GM_\odot}{R^3}\left[-(\alpha+\tfrac{1}{2}\gamma)+\frac{3}{2}\frac{X^2+Y^2}{R^2}\alpha+\frac{3}{2}\frac{Z^2}{R^2}\gamma\right], \tag{4.234}$$

where M is the earth's mass and

$$\alpha = \int \rho x^2 \mathrm{d}^3\boldsymbol{x} = \int \rho y^2 \mathrm{d}^3\boldsymbol{x}, \qquad \gamma = \int \rho z^2 \mathrm{d}^3\boldsymbol{x}. \tag{4.235}$$

Using the relations

$$A = B = \alpha+\gamma, \qquad C = 2\alpha, \tag{4.236}$$

we get

$$U = -\frac{GMM_{\odot}}{R} - GM_{\odot}(C-A)\frac{X^2+Y^2-2Z^2}{2R^5}. \tag{4.237}$$

The first term will not give a contribution to $\mathcal{M}$, and from the second term we get by using (4.231), (4.232), and (4.237)

$$\begin{aligned} \mathcal{M}_{\xi} &= \frac{3GM_{\odot}(C-A)YZ}{R^5}, \\ \mathcal{M}_{\eta} &= \frac{3GM_{\odot}(A-C)XZ}{R^5}, \\ \mathcal{M}_{\zeta} &= 0. \end{aligned} \tag{4.238}$$

Let χ be the angle between ES and the ξ-axis (i.e., the sun's longitude; see fig. 22) and δ the angle between the ecliptic and the equatorial plane. We then get, using (4.101) (see fig. 23; this figure is similar to fig. 18, if in that figure

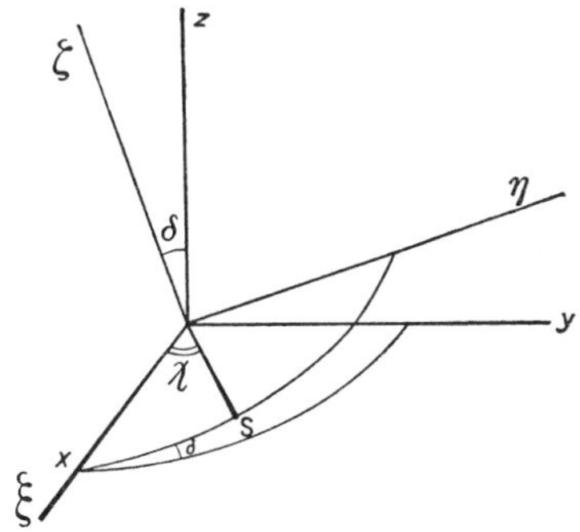

Fig. 23. The position of the sun in its orbit around the earth is given by the angles δ, the angle between the ecliptic plane and the earth's equatorial plane, and χ, the sun's longitude.

we put $\theta = \delta$, $\psi = \chi$, and $\varphi = 0$; it must, however, be realised that the XYZ-axes in fig. 18 are fixed in the rigid body, while here the ζ-axis is fixed in the earth, but the line ES rotates in the $\xi\eta$-plane, while the $\xi\eta$-axes themselves rotate with respect to the earth):

$$X = R\cos\chi, \qquad Y = R\sin\chi\cos\delta, \qquad Z = R\sin\chi\sin\delta. \tag{4.239}$$

Putting

$$K = 3GM_{\odot}(C-A)/R^3, \tag{4.240}$$

equation (4.229) can be written as follows:

$$\begin{aligned} A\dot{p}+Cr\omega_\eta-Aq\omega_\zeta &= K\sin^2\chi\sin\delta\cos\delta,\\ A\dot{q}+Ap\omega_\zeta-Cr\omega_\xi &= K\sin\chi\cos\chi\sin\delta,\\ C\dot{r}+Aq\omega_\xi-Ap\omega_\eta &= 0. \end{aligned} \tag{4.241}$$

To find expressions for the components of ω_0 we notice that the only difference between the $\xi\eta\zeta$-system and a system with axes fixed along the earth's principal axes is that the first one is rotating around the ζ-axis with respect to the second system. This means that

$$\omega_\xi = p, \qquad \omega_\eta = q, \qquad \omega_\zeta = r-\dot{\psi}, \tag{4.242}$$

that is, $\omega_\zeta \neq r$.

From the third of equation (4.241) and equation (4.242) it follows that

$$r = \text{constant} = \Omega, \tag{4.243}$$

where Ω is the angular velocity of the earth's daily rotation.

At this stage it is convenient to restrict ourselves to approximate solutions of equations (4.241). This is possible, if we remind ourselves that to a first approximation $\omega_\zeta = 0$, $p = 0$, and $q = 0$ (the position of the earth's axis with respect to the ecliptic is not changed very much) so that we can neglect terms of the second order in these quantities. In the same approximation we should be able to neglect the rate of change of p and q. Indeed, one can shew that $\dot{p}$ and $\dot{q}$ are smaller than q, or ω_ζ, by a factor of the order of magnitude of the period of the sun in its orbit around the earth to the period of rotation of the earth around its axis [see, e.g., A. G. Webster, The Dynamics of Particles, reprint of second edition (Dover, New York, 1959) § 96]. Neglecting all higher order terms we get from the first two equations (4.241)

$$\begin{aligned} p &= -(K/C\Omega)\sin\chi\cos\chi\sin\delta,\\ q &= (K/C\Omega)\sin^2\chi\sin\delta\cos\delta. \end{aligned} \tag{4.244}$$

From the relationship between figs. 18 and 23 and equations (4.103) it follows that, putting $\psi = 0$, $\theta = \delta$

$$\begin{aligned} \dot{\delta} &= -(K/C\Omega)\sin\chi\cos\chi\sin\delta,\\ \dot{\varphi} &= (K/C\Omega)\sin^2\chi\cos\delta. \end{aligned} \tag{4.245}$$

To a fair approximation we can write for the sun's longitude χ,

$$\chi = \lambda_0+\omega t, \tag{4.246}$$

where ω corresponds to a period of one year. We notice that the time average of $\dot{\delta}$ vanishes: there is no secular contribution to the nutation of the earth's axis in our present approximation. For the secular precession we get the equation

$$\overline{\dot{\varphi}} = K \cos \delta / 2C\Omega, \tag{4.247}$$

where we have replaced $\sin^2 \chi$ by its time average, $\frac{1}{2}$.

So far we have only taken the sun's influence into account. A similar calculation can be made for the moon's influence. As the moon's orbit is practically in the plane of the ecliptic, we can to a first approximation simply replace K by K', where K' is obtained from (4.240) with $M_\odot$ and R replaced by the lunar mass and the distance from the moon to the earth. Although the moon's mass is much smaller than the sun's mass, it is so much nearer to the earth that the value of K' is just over twice that of K.

To find the period of the precession we write

$$\frac{K \cos \delta}{2C\Omega} = \frac{GM_\odot}{R^3} \frac{1}{\Omega} \frac{C-A}{C} \frac{3 \cos \delta}{2} = \frac{3 \cos \delta}{2} \frac{\omega}{\Omega} \frac{C-A}{C} \omega \doteqdot \frac{1}{80000} \omega, \tag{4.248}$$

where we have used Kepler's third law (1.246) and the values $C/(C-A) \doteqdot 300$, $\Omega/\omega \doteqdot 365$, $\delta \doteqdot 23°$. The difference between this period of about 80 000 years and the observed one of about 26 000 years is readily explained by taking the influence of the moon into account.

3. ROTATING FRAMES OF REFERENCE; THE CORIOLIS FORCE

In the preceding section we derived the Euler equations, using the relation (4.203) between rates of change in a frame of reference fixed in space and rates of change in a rotating frame of reference. In the present section we shall apply (4.203) to (4.107) instead of to (4.112). We shall especially consider the motion of point particles on the earth. If we are dealing with a particle of mass m on which a force $\boldsymbol{F}$ is acting, the equation of motion is

$$\left(\frac{\mathrm{d}^2 \boldsymbol{x}}{\mathrm{d}t^2}\right)_{xyz} = \frac{\boldsymbol{F}}{m}, \tag{4.301}$$

and using (4.203) twice, we find in the rotating frame of reference

$$m\ddot{\boldsymbol{x}} = -2m[\boldsymbol{\omega} \wedge \dot{\boldsymbol{x}}] - m\{[\dot{\boldsymbol{\omega}} \wedge \boldsymbol{x}] + [\boldsymbol{\omega} \wedge [\boldsymbol{\omega} \wedge \boldsymbol{x}]]\} + \boldsymbol{F}, \tag{4.302}$$

where a dot indicates $(\mathrm{d}/\mathrm{d}t)_{XYZ}$.

In the cases to be considered here ω will be the rotational vector corresponding to the earth's rotation. To a very high degree of accuracy we can thus take $\dot{\omega} = 0$. The expression $[\omega \wedge [\omega \wedge x]]$ is easily seen to be a centrifugal acceleration. Its component along the radius vector from the earth's centre to a point on the earth's surface, which is equal to $\Omega^2 R \cos \varphi$ (Ω: rotational velocity of the earth around its axis, R: earth's radius; φ latitude of the point on the earth's surface), is about $0.003g \cos \varphi$ (g: gravitational acceleration) and can be taken into account together with the ordinary gravitational acceleration. It is a correction of at most 0.3 per cent to g, and we shall not consider it any further in the following. We shall also neglect the other components of $[\omega \wedge [\omega \wedge x]]$.

We are now left with the following (approximate) equation of motion

$$m\ddot{x} = -2m[\omega \wedge \dot{x}] + F; \tag{4.303}$$

the expression on the right hand side is the so-called *Coriolis force*. If a particle is moving in the Northern hemisphere in a northerly direction, it will be subject to a force in an easterly direction. This is of importance for an understanding of the course taken by rivers and for an understanding of the formation and sense of circulation of cyclones. We shall restrict ourselves to two simple mechanical examples where the Coriolis force plays a role. The first one is the motion of a particle in the gravitational field of the earth, that is, the path of a point particle dropped from a height; the second one is the case of the Foucault pendulum.

We shall choose our axes as follows: the z-axis will be in the direction of the radius vector from the centre of the earth to the point on the surface of the earth where we perform the experiment; the y-axis is taken along the circle of constant latitude in the direction W → E, and the x-axis is then taken along the meridian in the N → S direction. (We note that the directions of the xyz-axes are different in different points on earth.) The vector ω has in this system of axes the components $-\Omega \cos \varphi$, 0, $\Omega \sin \varphi$ where Ω is again the earth's rotational velocity and φ the latitude of the point under consideration. Equation (4.303) then becomes

$$\begin{aligned} \ddot{x} &= 2\Omega \sin\varphi\, \dot{y} + F_x/m, \\ \ddot{y} &= -2\Omega \cos\varphi\, \dot{z} - 2\Omega \sin\varphi\, \dot{x} + F_y/m, \\ \ddot{z} &= 2\Omega \cos\varphi\, \dot{y} + F_z/m. \end{aligned} \tag{4.304}$$

For the first case we can put $F_x = F_y = 0$, $F_z = -mg$ (where g can be corrected for the centrifugal contribution mentioned a moment ago). We can integrate the first and the third of equations (4.304) and substitute the result into the second equation (4.304) (we use the boundary conditions $x = y = 0$, $z = h$, and $\dot{x} = \dot{y} = \dot{z} = 0$ at $t = 0$):

$$\ddot{y} = -4\Omega^2 y + 2\Omega \cos \varphi \, gt. \tag{4.305}$$

We can neglect the term in Ω^2 with respect to the other terms, and we find for the displacement Δy of the point where the mass hits the earth from the point where it would hit, if Ω were equal to zero:

$$\Delta y = \tfrac{1}{3}\Omega \cos \varphi \, gT^3, \tag{4.306}$$

where T is the time it takes the particle to reach the ground,

$$T \doteqdot (2h/g)^{\frac{1}{2}}. \tag{4.307}$$

Substituting (4.307) into (4.306) and taking $h = 10^2$ m, we find a value of about 0.02 m for Δy! The physical reason for this deviation is simple: the mass was dropped at a point where the W → E velocity was greater than at the ground!

In the second case we are dealing with a pendulum with two degrees of freedom. The most rigorous method to treat this problem is to consider it to be a spherical pendulum and to use the Lagrangian equations of motion of the second kind. From the geometry of the problem we see, however, that $\dot{z}$ will be a quantity of the second order, if we use a small vibrations approximation (z will differ from its equilibrium position only by an amount quadratic in the amplitude). Moreover, from the theory of the simple pendulum, we expect F_x/m, and F_y/m to be, respectively, equal to $-gx/l$ and $-gy/l$, where l is the length of the pendulum. Neglecting $\dot{z}$ in the second of equations (4.304), the first two equations (4.304) are

$$\begin{aligned} \ddot{x} - 2\Omega \sin \varphi \, \dot{y} &= -gx/l, \\ \ddot{y} + 2\Omega \sin \varphi \, \dot{x} &= -gy/l. \end{aligned} \tag{4.308}$$

We notice that (4.308) reduce to the usual simple pendulum equations if $\Omega = 0$.

The solution of (4.308) is a simple, harmonic, pendulum motion, but with the plane of the oscillation rotating with an angular velocity $\Omega \sin \varphi$. This can be seen immediately, if we introduce an $x'y'$-system which rotates with an angular velocity $\Omega \sin \varphi$ with respect to the xy-system. If one does this, the terms proportional to Ω vanish. Put differently: these terms are the two-dimensional analogues of the Coriolis acceleration.

Analytically we find the solution by introducing a complex variable:

$$u = x + \mathrm{i}y. \tag{4.309}$$

In this variable we have

$$\ddot{u} + 2\mathrm{i}\Omega \sin \varphi \dot{u} + gu/l = 0. \tag{4.310}$$

If we try a solution of the form

$$u = A\mathrm{e}^{\mathrm{i}\omega t}, \tag{4.311}$$

we find, neglecting terms in Ω^2,

$$\omega = \pm\omega_0 - \Omega \sin \varphi, \qquad \omega_0 = (g/l)^{\frac{1}{2}}, \tag{4.312}$$

or,

$$u = (A\mathrm{e}^{\mathrm{i}\omega_0 t} + B\mathrm{e}^{-\mathrm{i}\omega_0 t})\mathrm{e}^{-\mathrm{i}\Omega \sin \varphi t}. \tag{4.313}$$

The expression within the brackets represents an elliptic orbit (see fig. 24):

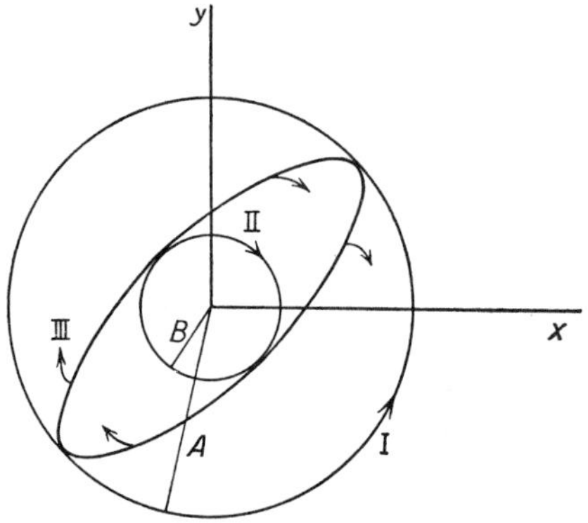

Fig. 24. Foucault's pendulum experiment. The ellipse III is the actual path described by the end of the pendulum.

$A\mathrm{e}^{\mathrm{i}\omega_0 t}$ represents the motion in the circle I, $B\mathrm{e}^{-\mathrm{i}\omega_0 t}$ the motion in the circle II, the ellipse III is the combination of the two, and the factor $\mathrm{e}^{-\mathrm{i}\Omega \sin \varphi t}$ means that this ellipse rotates clockwise with an angular velocity $\Omega \sin \varphi$.

This rotation of the plane of oscillation of a pendulum was shewn experimentally by Foucault in his famous 1851 experiment in the Panthéon, for which he received the Copley Medal of the Royal Society in 1855.

CHAPTER 5

THE CANONICAL EQUATIONS OF MOTION

In this chapter we first of all shew how one can describe the motion of a mechanical system with s degrees of freedom in the $2s$-dimensional phase space. The canonical equations of motion are derived from the Lagrangian equations of motion. A brief discussion is given of canonical transformations, and a longer one of the properties of the Poisson brackets, their invariance under canonical transformations, their usefulness in finding integrals of motion, and their connexion with infinitesimal contact transformations. The case of a particle in an electromagnetic field is briefly discussed. In the last section the principle of least action is derived from Hamilton's variational principle and we discuss how the time can be introduced on an equivalent footing with the q_k.

1. THE HAMILTONIAN EQUATIONS OF MOTION.

We have seen in the last few chapters that the Lagrangian equations of motion are in many cases a very appropriate way of describing the behaviour of mechanical systems. The Lagrangian equations of motion are a set of s second order differential equations. In many cases it is more convenient to have a set of $2s$ first order differential equations. In the Lagrangian function $L(q_k, \dot{q}_k)$, the q_k and $\dot{q}_k$ are not independent variables, since the $\dot{q}_k$ are the time derivatives of the q_k. The easiest way to remedy this is to introduce s new variables, r_k, through the equations

$$\dot{q}_k = r_k. \tag{5.101}$$

Equations (5.101) are s first order differential equations connecting the $2s$ variables q_k and r_k. The Lagrangian is now a function $L(q_k, r_k)$ of the same $2s$ variables and the Lagrangian equations of motion (2.308) are now also first order equations,

$$\frac{\mathrm{d}}{\mathrm{d}t}\frac{\partial L}{\partial r_k} - \frac{\partial L}{\partial q_k} = 0. \tag{5.102}$$

Equations (5.101) and (5.102) together form a set of $2s$ first order differential equations and our task is completed. As before, the state of a mechanical system is completely determined, once we have given the values of the q_k and of the r_k at one particular time. However, while in Chapter 2 we considered an orbit of the system to be a curve in the s-dimensional q-space, now an orbit will be represented by a curve in the $2s$-dimensional q, r-space.

It is convenient to use instead of the q_k, r_k another set of variables in terms of which the equations of motion are more symmetrical than equations (5.101) and (5.102). To do this we introduce again the generalised momenta, p_k, by equation (2.310), which we can now write as

$$p_k = \frac{\partial L}{\partial r_k}. \tag{5.103}$$

This equation, or rather, these s equations together with the trivial equations $q_k = q_k$ are the equations through which we transform from the q_k, r_k to the q_k, p_k. To find the equations of motion in terms of the q_k, p_k, we introduce instead of the Lagrangian $L(q_k, r_k)$ the *Hamiltonian* of the system by the equation

$$H = \sum_k p_k r_k - L. \tag{5.104}$$

We now consider the variation of H,

$$\delta H = \sum_k p_k \delta r_k + \sum_k r_k \delta p_k - \delta L, \tag{5.105}$$

or, if we use the equations of motion (5.101) and (5.102) and the definition (5.103), so that

$$\delta L = \sum_k \frac{\partial L}{\partial q_k} \delta q_k + \sum_k \frac{\partial L}{\partial r_k} \delta r_k = \sum_k \dot{p}_k \delta q_k + \sum_k p_k \delta r_k, \tag{5.106}$$

we have

$$\delta H = \sum_k r_k \delta p_k - \sum_k \dot{p}_k \delta q_k = \sum_k \dot{q}_k \delta p_k - \sum_k \dot{p}_k \delta q_k. \tag{5.107}$$

It follows thus that

$$\dot{q}_k = \frac{\partial H}{\partial p_k}, \qquad \dot{p}_k = -\frac{\partial H}{\partial q_k}. \tag{5.108}$$

Equations (5.108) are the *Hamiltonian equations of motion* or the *canonical equations of motion.*

In deriving equations (5.108) we have used as long as possible the q_k, r_k

set of coordinates, to stress the fact that we are dealing with a set of $2s$ independent variables. In many accounts of the theory of Lagrangian and Hamiltonian equations of motion, this fact is not always emphasised, and instead of (5.104), (5.105), and (5.106) one finds

$$H = \sum_k p_k \dot{q}_k - L, \tag{5.104'}$$

$$\delta H = \sum_k p_k \delta \dot{q}_k + \sum_k \dot{q}_k \delta p_k - \delta L, \tag{5.105'}$$

$$\delta L = \sum_k \frac{\partial L}{\partial q_k} \delta q_k + \sum_k \frac{\partial L}{\partial \dot{q}_k} \delta \dot{q}_k = \sum_k \dot{p}_k \delta q_k + \sum_k p_k \delta \dot{q}_k. \tag{5.106'}$$

In fact we shall use these equations when discussing the Hamiltonian formalism for continuous media in chapter 8. Moreover, we have assumed throughout that L does not depend explicitly on the time; this is then also true for H.

We wish to draw attention to the resemblance between (5.104), (5.107), and (5.108) on the one hand, and (2.404) (or (2.410)), (2.406), and (2.407) and (2.409), on the other hand.

To find the physical meaning of H we shall once again assume that the kinetic energy T is a homogeneous quadratic expression in the $\dot{q}_k$ and that the potential energy U is independent of the $\dot{q}_k$. We then have from Euler's relation for homogeneous polynomials,

$$2T = \sum_k \frac{\partial T}{\partial \dot{q}_k} \dot{q}_k = \sum_k \frac{\partial L}{\partial \dot{q}_k} \dot{q}_k = \sum_k p_k \dot{q}_k, \tag{5.109}$$

and from (5.104') and (2.235)

$$H = \sum_k p_k \dot{q}_k - L = 2T - (T - U) = T + U, \tag{5.110}$$

which shews that H is the total energy of the system, expressed in terms of the p_k and q_k.

If H does not contain the time explicitly, it follows immediately from (5.108) that the energy is an integral of motion,

$$\frac{dH}{dt} = \sum_k \frac{\partial H}{\partial p_k} \dot{p}_k + \sum_k \frac{\partial H}{\partial q_k} \dot{q}_k + \frac{\partial H}{\partial t} = \frac{\partial H}{\partial t} = 0. \tag{5.111}$$

One might inquire as to the importance of the canonical equations of motion. We may mention here two aspects. The first one is that quantum mechanics, both the old quantum mechanics and modern wave or matrix mechanics, is based upon the Hamiltonian formalism rather than on the Lagrangian one, although the Lagrangian formalism is useful in developing field theory. The second one is that the Hamiltonian formalism is especially suited to develop perturbation theory, that is, to consider those systems for which we cannot obtain an exact solution of the equations of motion. As these systems are the rule rather than the exception, it is clear that there is ample scope for perturbation theory — both in classical and in quantum mechanics. We shall return to perturbation theory in Chapter 7, but in the remainder of this chapter and in the next chapter we shall develop the formalism which will be needed to deal with perturbations. Finally, we may mention the fact that statistical mechanics uses the Hamiltonian approach; the $2s$-dimensional p, q-space is there called *phase space.*

From the fact that the Lagrangian equations of motion are invariant under a transformation from one set of q_k to another set q'_k [see (2.309)], it follows immediately that, provided we define the p'_k by the equations

$$p'_k = \frac{\partial L(q', \dot{q}')}{\partial \dot{q}'_k}, \tag{5.112}$$

the equations of motion in terms of the q'_k and p'_k will be

$$\dot{q}'_k = \frac{\partial H(p', q')}{\partial p'_k}, \qquad \dot{p}'_k = -\frac{\partial H(p', q')}{\partial q'_k}. \tag{5.113}$$

Once we have introduced sets of $2s$ coordinates to describe our system, there are, however, more general transformations from one set of coordinates to another set, and these will be considered in the next section.

2. CANONICAL TRANSFORMATIONS

In many cases it is impossible to solve the equations of motion (5.108). One way to simplify the equations may be through a transformation from the p_k, q_k to another set of variables α_k, β_k ($k = 1, 2, \ldots, s$), say. If the equations of motion are simpler in the new variables than in the old ones, we have a clear gain. We shall not consider all possible transformations, but only the so-called *canonical* or *contact transformations*, which are transformations such that in the new variables the equations of motion are again of the canonical form. Thus, if

$$p_k = p_k(\alpha, \beta), \qquad q_k = q_k(\alpha, \beta) \tag{5.201}$$

is a canonical transformation from the p, q set to an α, β set, the equations of motion in the α, β will be

$$\dot{\alpha}_k = -\frac{\partial \overline{H}}{\partial \beta_k}, \qquad \dot{\beta}_k = \frac{\partial \overline{H}}{\partial \alpha_k}, \tag{5.202}$$

where $\overline{H}$ is the Hamiltonian (energy) expressed in the α, β.

We can easily find some simple examples of canonical transformations. For instance, the transformations

$$q_k = \alpha_k, \qquad p_k = -\beta_k,$$

or

$$q_k = -\alpha_k, \qquad p_k = \beta_k,$$

clearly are canonical.

Also, the point transformation

$$\beta_k = \beta_k(q_1, \ldots, q_s), \qquad \alpha_k = \frac{\partial L}{\partial \dot{\beta}_k} \tag{5.203}$$

is a canonical one, as we discussed at the end of the previous section.

We shall now shew that a necessary and sufficient condition for a transformation from the p_k, q_k to the α_k, β_k to be a canonical transformation is that there exists a function $W(q_k, \beta_k)$ such that

$$p_k = \frac{\partial W}{\partial q_k}, \qquad \alpha_k = -\frac{\partial W}{\partial \beta_k}. \tag{5.204}$$

First of all, it follows from the principles of the theory of variations that

$$\delta \frac{\mathrm{d}W}{\mathrm{d}t} - \frac{\mathrm{d}}{\mathrm{d}t}\delta W = 0, \tag{5.205}$$

which reduces to the relation

$$-\sum_k \dot{p}_k \delta q_k + \sum_k \dot{q}_k \delta p_k + \sum_k \dot{\alpha}_k \delta \beta_k - \sum_k \dot{\beta}_k \delta \alpha_k = 0. \tag{5.206}$$

From equations (5.108) it follows that

$$-\sum_k \dot{p}_k \delta q_k + \sum_k \dot{q}_k \delta p_k = \delta H, \tag{5.207}$$

and combining (5.206) and (5.207) we have

$$\delta H = \delta \overline{H} = -\sum_k \dot{\alpha}_k \delta \beta_k + \sum_k \dot{\beta}_k \delta \alpha_k, \tag{5.208}$$

from which equations (5.202) follow immediately, as H and $\overline{H}$ are the same function, but expressed in different sets of coordinates.

We must now shew that if the transformation is canonical, we can find a function $W(q, \beta)$ such that equations (5.204) hold. These equations are, of course, the equations determining the transformation.

To find the function $W(q, \beta)$ we first shew how the canonical equations of motion follow from a modified Hamiltonian principle, namely, the condition that

$$\int L\,\mathrm{d}t = \text{extremum}, \tag{5.209}$$

where now, in contradistinction to the original condition (2.234) the integrand is considered as a function of $2s$ variables. Using again the r_k to emphasise this fact we consider the extremum problem

$$\delta \int L(q_k, r_k)\mathrm{d}t = 0 \tag{5.210}$$

with (5.101) as subsidiary conditions.

We saw in Chapter 2 how one uses the method of Lagrangian multipliers to deal with a variational problem which involves subsidiary conditions. As the present case has some complicating features, we shall discuss the steps to be taken in some detail. If we write out (5.210) in detail, we get under the integral sign the variations δq_k and δr_k which are now functions of t which should therefore have been indicated by writing explicitly $\delta q_k(t)$ and $\delta r_k(t)$. These functions of t are not independent but they must satisfy the conditions

$$\delta \dot{q}_k(t) - \delta r_k(t) = 0, \tag{5.211}$$

where, as in Chapter 2, the $\delta \dot{q}_k(t)$ are the time derivatives of the $\delta q_k(t)$. In the usual way one can now rid oneself of the restrictions on the variations δq_k and δr_k by multiplying each of the s left hand sides of equations (5.211) by factor λ_k, which is a function of time, and adding the resulting expressions to the integrand in (5.210). One finds thus the new extremum problem

$$\delta \int [L + \sum_k \lambda_k(\dot{q}_k - r_k)]\mathrm{d}t = 0. \tag{5.212}$$

The variations of the q_k and of the r_k can now be considered to be independent, and it follows from a consideration of the coefficients of the

δr_k that

$$\lambda_k = \frac{\partial L}{\partial r_k}, \tag{5.213}$$

so that we get from (5.212)

$$\delta \int \left[L + \sum_k \frac{\partial L}{\partial r_k} (\dot{q}_k - r_k) \right] \mathrm{d}t = 0. \tag{5.214}$$

Using (5.103) and (5.104) we see that we are led to the variational principle

$$\delta \int (\sum_k p_k \dot{q}_k - H) \mathrm{d}t = 0, \tag{5.215}$$

where the integrand is a function of the 2s variables p_k and q_k, and where the $\dot{q}_k$ are functions of the same variables through the equations

$$p_k = \frac{\partial L}{\partial \dot{q}_k}. \tag{5.216}$$

Writing down explicitly the variation of the integrand in (5.215), integrating the terms involving $\delta \dot{q}_k$ by parts, and taking into account that by virtue of (2.306) the integrated terms vanish, we get

$$\int \sum_k \left[\left(\dot{q}_k - \frac{\partial H}{\partial p_k} \right) \delta p_k - \left(\dot{p}_k + \frac{\partial H}{\partial q_k} \right) \delta q_k \right] \mathrm{d}t = 0, \tag{5.217}$$

from which equations (5.108) follow. We see therefore that the variational principle (5.215) is equivalent to the canonical equations of motion (5.108).

If the transformation from the p_k, q_k to the α_k, β_k is canonical, (5.215) must lead to a similar equation in terms of the α_k, β_k,

$$\delta \int (\sum_k \alpha_k \dot{\beta}_k - \overline{H}) \mathrm{d}t = 0, \tag{5.218}$$

where $\overline{H}$ is the same as H, but now expressed in terms of the α_k, β_k. Equations (5.215) and (5.218) can hold at the same time, only if

$$\sum_k p_k \dot{q}_k - H = \sum_k \alpha_k \dot{\beta}_k - \overline{H} + \frac{\mathrm{d}}{\mathrm{d}t} W(q, \beta). \tag{5.219}$$

As $H = \overline{H}$, it follows that equations (5.204) hold which concludes our proof.

The function $W(q, \beta)$ is called a *generating function*: the equations (5.204) generate the canonical transformation. It is possible to consider other generating functions. In fact, there are four different combinations of two sets of s variables: q_k, β_k; q_k, α_k; p_k, β_k; and p_k, α_k which seem to be the obvious ones to take. The four possible generating functions, and the corresponding equations determining the transformation are the following ones:

$$\begin{aligned}
&\text{(a)}: \quad W(q_k, \beta_k); \quad p_k = \frac{\partial W}{\partial q_k}, \quad \alpha_k = -\frac{\partial W}{\partial \beta_k}; \quad W; \\
&\text{(b)}: \quad S(q_k, \alpha_k); \quad p_k = \frac{\partial S}{\partial q_k}, \quad \beta_k = \frac{\partial S}{\partial \alpha_k}; \quad W = S - \sum_k \alpha_k \beta_k; \\
&\text{(c)}: \quad T(p_k, \beta_k); \quad q_k = -\frac{\partial T}{\partial p_k}, \quad \alpha_k = -\frac{\partial T}{\partial \beta_k}; \quad W = T + \sum_k p_k q_k; \\
&\text{(d)}: \quad U(p_k, \alpha_k); \quad q_k = -\frac{\partial U}{\partial p_k}, \quad \beta_k = \frac{\partial U}{\partial \alpha_k}; \quad W = U + \sum_k p_k q_k - \sum_k \alpha_k \beta_k.
\end{aligned} \tag{5.220}$$

We have also given the relation between the generating function and the function W which, according to the theorem we proved earlier in this section, should always exist for any canonical transformation.

The transformations defined by (5.220) are examples of *Legendre transformations* — which play such an important role in thermodynamics.

Let us consider a few very simple transformations:

$$\begin{aligned}
&\text{(i)} \quad W = \sum_k q_k \beta_k; \\
&\text{(ii)} \quad T = -\sum_k p_k \beta_k; \\
&\text{(iii)} \quad U = \sum_k p_k \alpha_k; \\
&\text{(iv)} \quad S = \sum_k \alpha_k Q_k(q).
\end{aligned} \tag{5.221}$$

The first transformation leads to

$$p_k = \beta_k, \qquad q_k = -\alpha_k; \tag{5.222}$$

this transformation is the same as the one considered briefly at the beginning of this section; it leads to a set of variables where the old momenta are the new 'coordinates' and the old coordinates the new 'momenta'. This shews that there is little justification in continuing to use the nomenclature 'momenta' and 'coordinates' for the α_k and β_k. It is better to call them *canonically conjugate variables.*

We may draw attention to the fact that the transformation (5.222) is one where there does not exist a generating function T. From (5.220) and (5.222) it follows easily that for this case $T \equiv 0$.

The second transformation is the identical one:

$$q_k = \beta_k, \qquad \alpha_k = p_k. \tag{5.223}$$

In this case we find that $W \equiv 0$. It is the same as the transformation which would be generated by

$$S = \sum_k \alpha_k q_k. \tag{5.224}$$

The third transformation

$$q_k = -\alpha_k, \qquad \beta_k = p_k, \tag{5.225}$$

is the same as the first one.

The last transformation is a point transformation, like the transformation (5.203),

$$\beta_k = Q_k(q_k), \qquad p_k = \sum_l \alpha_l \frac{\partial Q_l}{\partial q_k}. \tag{5.226}$$

We do not wish to consider at this moment any other transformations as we shall consider a large class of transformations of the type (5.220b) in the next chapter.

3. POISSON AND LAGRANGIAN BRACKETS; INFINITESIMAL TRANSFORMATIONS

The canonical equations of motion (5.108) describe the behaviour of the p_k and the q_k. From these equations we can find the equation of motion for any function $F(p_k, q_k)$ of the p_k, q_k (we shall assume for the sake of simplicity that F does not contain the time explicitly). It follows easily that

$$\dot{F} = \sum_k \left(\frac{\partial F}{\partial q_k}\dot{q}_k + \frac{\partial F}{\partial p_k}\dot{p}_k\right) = \sum_k \left(\frac{\partial F}{\partial q_k}\frac{\partial H}{\partial p_k} - \frac{\partial F}{\partial p_k}\frac{\partial H}{\partial q_k}\right), \tag{5.301}$$

where we have used equations (5.108) and the fact that in accordance with our assumptions $\partial F/\partial t = 0$.

If we introduce the notation

$$\{f, g\} = \sum_k \left(\frac{\partial f}{\partial q_k}\frac{\partial g}{\partial p_k} - \frac{\partial f}{\partial p_k}\frac{\partial g}{\partial q_k}\right) = -\{g, f\}, \tag{5.302}$$

we can write (5.301) in the form

$$\dot{F} = \{F, H\}. \tag{5.303}$$

The $\{f, g\}$ are called *Poisson brackets*. The advantage of the form (5.303) is that it does not depend on the choice of coordinates since the Poisson brackets are invariant under a canonical transformation as we shall see presently. In the present section we shall consider some properties of the Poisson brackets and also of the related *Lagrangian brackets* which are defined by the equation

$$[f, g] = \sum_k \left(\frac{\partial q_k}{\partial f}\frac{\partial p_k}{\partial g} - \frac{\partial p_k}{\partial f}\frac{\partial q_k}{\partial g}\right) = -[g, f]. \tag{5.304}$$

The Lagrangian brackets are also invariant under contact transformations. From a comparison of (5.302) and (5.304) one sees that in a certain sense the Lagrangian brackets are the inverse of the Poisson brackets. This statement can be made more precise as follows. If γ_k $(k = 1, 2, \ldots, 2s)$ is a set of $2s$ independent functions of the p_k and q_k, one can prove by a straightforward substitution that

$$\sum_k \{\gamma_k, \gamma_l\}[\gamma_k, \gamma_m] = \delta_{lm}, \tag{5.305}$$

where δ_{kl} is again the Kronecker delta-function defined by (3.132).

We shall now consider the so-called fundamental brackets, that is, the brackets concerning the p_k and q_k. As the p_k and q_k are independent variables, we have

$$\frac{\partial p_k}{\partial p_l} = \delta_{kl}, \qquad \frac{\partial p_k}{\partial q_l} = 0, \qquad \frac{\partial q_k}{\partial p_l} = 0, \qquad \frac{\partial q_k}{\partial q_l} = \delta_{kl}. \tag{5.306}$$

From (5.302), (5.304), and (5.306) it follows easily that

$$\{q_k, p_l\} = \delta_{kl}, \qquad \{q_k, q_l\} = 0, \qquad \{p_k, p_l\} = 0; \tag{5.307}$$

$$[q_k, p_l] = \delta_{kl}, \qquad [q_k, q_l] = 0, \qquad [p_k, p_l] = 0. \tag{5.308}$$

Equations (5.307) and (5.308) are an invariant property of canonical variables. Indeed, evaluating $\partial \bar{H}(\alpha, \beta)/\partial \alpha_i$ we get

$$\begin{aligned}\frac{\partial \bar{H}}{\partial \alpha_i} &= \sum_k \left[\frac{\partial H}{\partial p_k}\frac{\partial p_k}{\partial \alpha_i} + \frac{\partial H}{\partial q_k}\frac{\partial q_k}{\partial \alpha_i}\right] \\ &= \sum_k \left[\dot{q}_k \frac{\partial p_k}{\partial \alpha_i} - \dot{p}_k \frac{\partial q_k}{\partial \alpha_i}\right] \\ &= \sum_{k,j} \left\{\left[\frac{\partial p_k}{\partial \alpha_i}\frac{\partial q_k}{\partial \beta_j} - \frac{\partial q_k}{\partial \alpha_i}\frac{\partial p_k}{\partial \beta_j}\right]\dot{\beta}_j + \left[\frac{\partial p_k}{\partial \alpha_i}\frac{\partial q_k}{\partial \alpha_j} - \frac{\partial q_k}{\partial \alpha_i}\frac{\partial p_k}{\partial \alpha_j}\right]\dot{\alpha}_j\right\} \\ &= \sum_j [\beta_j, \alpha_i]\dot{\beta}_j + \sum_j [\alpha_j, \alpha_i]\dot{\alpha}_j, \end{aligned} \tag{5.309}$$

from which by comparing (5.309) with (5.202) it follows that

$$[\beta_j, \alpha_i] = \delta_{ij}, \qquad [\alpha_j, \alpha_i] = 0. \tag{5.310}$$

The last fundamental Lagrangian bracket $[\beta_j, \beta_i]$ also vanishes as follows from considering $\partial \bar{H}/\partial \beta_i$.

Similarly we have for the α_k, β_k the relations

$$\{\beta_j, \alpha_i\} = \delta_{ij}, \qquad \{\alpha_i, \alpha_j\} = 0, \qquad \{\beta_i, \beta_j\} = 0. \tag{5.311}$$

To prove this, we consider $\dot{\alpha}_i$,

$$\begin{aligned}\dot{\alpha}_i &= \sum_k \left[\frac{\partial \alpha_i}{\partial q_k}\dot{q}_k + \frac{\partial \alpha_i}{\partial p_k}\dot{p}_k\right] \\ &= \sum_k \left[\frac{\partial \alpha_i}{\partial q_k}\frac{\partial H}{\partial p_k} - \frac{\partial \alpha_i}{\partial p_k}\frac{\partial H}{\partial q_k}\right] \\ &= \sum_{k,j} \left\{\left[\frac{\partial \alpha_i}{\partial q_k}\frac{\partial \alpha_j}{\partial p_k} - \frac{\partial \alpha_i}{\partial p_k}\frac{\partial \alpha_j}{\partial q_k}\right]\frac{\partial H}{\partial \alpha_j} + \left[\frac{\partial \alpha_i}{\partial q_k}\frac{\partial \beta_j}{\partial p_k} - \frac{\partial \alpha_i}{\partial p_k}\frac{\partial \beta_j}{\partial q_k}\right]\frac{\partial H}{\partial \beta_j}\right\} \\ &= \sum_j \{\alpha_i, \alpha_j\}\frac{\partial H}{\partial \alpha_j} + \sum_j \{\alpha_i, \beta_j\}\frac{\partial H}{\partial \beta_j}. \end{aligned} \tag{5.312}$$

Comparing (5.202) and (5.312) (and a similar equation for $\dot{\beta}_i$) equations (5.311) follow.

We have proved that the fundamental brackets satisfy (5.307) and (5.308) independent of our choice of canonical variables. This means that on going over from the p_k, q_k to the α_i, β_i the following equations hold,

$$\{q_k, p_l\}' = \{q_k, p_l\} = \delta_{kl}, \quad \{q_k, q_l\}' = \{q_k, q_l\} = 0, \quad \{p_k, p_l\}' = \{p_k, p_l\} = 0. \tag{5.313}$$

Strictly speaking we proved this for the α_i, β_i, but the same proof holds, of course, also when we consider the brackets of (5.313). In (5.313) we have indicated the Poisson brackets in terms of the α_i, β_i by a prime. Equations similar to (5.313) also hold for the Lagrangian brackets.

We shall now shew that the brackets — which up to now have been defined in terms of a particular set of canonical variables p_k, q_k — are invariant under a canonical transformation. We shall prove it for the Poisson brackets, that is, we shall prove

$$\{f, g\}' = \{f, g\}; \tag{5.314}$$

the proof for the Lagrangian brackets follows easily by the same method, or by using the fact that (5.305) holds independently of the choice of canonical variables.

Before proceeding with the proof we note that

$$\{p_k, f\} = -\frac{\partial f}{\partial q_k}, \qquad \{q_k, f\} = \frac{\partial f}{\partial p_k}. \tag{5.315}$$

If we substitute H for f, and use (5.303), equations (5.315) reduce to the canonical equations of motion (5.108). We now consider $\{f, g\}'$:

$$\begin{aligned}
\{f, g\}' &= \sum_i \left(\frac{\partial f}{\partial \beta_i}\frac{\partial g}{\partial \alpha_i} - \frac{\partial f}{\partial \alpha_i}\frac{\partial g}{\partial \beta_i}\right) \\
&= \sum_{i,k}\left[\frac{\partial f}{\partial \beta_i}\left(\frac{\partial g}{\partial q_k}\frac{\partial q_k}{\partial \alpha_i} + \frac{\partial g}{\partial p_k}\frac{\partial p_k}{\partial \alpha_i}\right) - \frac{\partial f}{\partial \alpha_i}\left(\frac{\partial g}{\partial q_k}\frac{\partial q_k}{\partial \beta_i} + \frac{\partial g}{\partial p_k}\frac{\partial p_k}{\partial \beta_i}\right)\right] \\
&= \sum_k \frac{\partial g}{\partial q_k}\{f, q_k\}' + \sum_k \frac{\partial g}{\partial p_k}\{f, p_k\}'.
\end{aligned} \tag{5.316}$$

Applying (5.316) first with $f \equiv q_l$ and then with $f \equiv p_l$, and using (5.313) and (5.315) we find

$$\begin{aligned}
\{q_l, g\}' &= \sum_k \frac{\partial g}{\partial q_k}\{q_l, q_k\}' + \sum_k \frac{\partial g}{\partial p_k}\{q_l, p_k\}' \\
&= \sum_k \frac{\partial g}{\partial q_k}\{q_l, q_k\} + \sum_k \frac{\partial g}{\partial p_k}\{q_l, p_k\} = \frac{\partial g}{\partial p_l} = \{q_l, g\};
\end{aligned} \tag{5.317}$$

$$\begin{aligned}
\{p_l, g\}' &= \sum_k \frac{\partial g}{\partial q_k}\{p_l, q_k\}' + \sum_k \frac{\partial g}{\partial p_k}\{p_l, p_k\}' \\
&= \sum_k \frac{\partial g}{\partial q_k}\{p_l, q_k\} + \sum_k \frac{\partial g}{\partial p_k}\{p_l, p_k\} = -\frac{\partial g}{\partial q_l} = \{p_l, g\}.
\end{aligned} \tag{5.318}$$

Having proved the invariance of the brackets $\{q_l, g\}$ and $\{p_l, g\}$, we get from (5.316), using equations (5.315),

$$\{f, g\}' = \sum_k \frac{\partial g}{\partial q_k}\{f, q_k\} + \sum_k \frac{\partial g}{\partial p_k}\{f, p_k\}$$

$$= \sum_k \left[\frac{\partial f}{\partial q_k}\frac{\partial g}{\partial p_k} - \frac{\partial f}{\partial p_k}\frac{\partial g}{\partial q_k}\right] = \{f, g\}, \qquad (5.319)$$

which completes the proof.

We shall now discuss constants of motion; for this discussion we need the so-called Jacobi identity,

$$\{f, \{g, h\}\} + \{g, \{h, f\}\} + \{h, \{f, g\}\} = 0. \qquad (5.320)$$

The proof of (5.320) is tedious but straightforward and will be left to the reader; one uses the relation

$$\{f, gh\} = g\{f, h\} + \{f, g\}h. \qquad (5.321)$$

From (5.303) it follows that a necessary and sufficient condition for F to be a constant of motion is that

$$\{F, H\} = 0. \qquad (5.322)$$

If we now use (5.320) with $f \equiv F, g \equiv G, h \equiv H$, where H is the Hamiltonian and F and G are constants of motion we find

$$\{H, \{F, G\}\} = -\{F, \{G, H\}\} + \{G, \{F, H\}\} = 0, \qquad (5.323)$$

or: if F and G are constants of motion, their Poisson bracket $\{F, G\}$ will also be a constant of motion. This is often a means of constructing new constants of motion, though not always, as we shall see presently.

As an example of constants of motion we may mention the total angular momentum and the total linear momentum of a system of particles. We shall discuss the properties of these quantities in some detail.

Let us first consider the angular momentum vector $\boldsymbol{M}$ of a system of particles. We shall assume that this system can be described in terms of Cartesian coordinates $\boldsymbol{x}_i$ so that the generalised momenta are the linear momenta $\boldsymbol{p}_i$. The total angular momentum vector is given by (1.309) which in terms of the $\boldsymbol{x}_i$ and $\boldsymbol{p}_i$ is of the form

$$\boldsymbol{M} = \sum_i [\boldsymbol{x}_i \wedge \boldsymbol{p}_i]. \qquad (5.324)$$

From (5.324) and (5.307) it follows easily that

$$\{M_x, M_y\} = M_z, \quad \{M_y, M_z\} = M_x, \quad \{M_z, M_x\} = M_y. \tag{5.325}$$

One can also prove the following relations

$$\{M^2, M_x\} = 0, \quad \{M^2, M_y\} = 0, \quad \{M^2, M_z\} = 0; \tag{5.326}$$

$$\begin{gathered} \{P_x, M_x\} = 0, \quad \{P_y, M_y\} = 0, \quad \{P_z, M_z\} = 0; \\ \{P_x, M^2\} = 0, \quad \{P_y, M^2\} = 0, \quad \{P_z, M^2\} = 0; \\ \{P_x, M_y\} = \{M_x, P_y\} = P_z, \quad \{P_y, M_z\} = \{M_y, P_z\} = P_x, \\ \{P_z, M_x\} = \{M_z, P_x\} = P_y; \end{gathered} \tag{5.327}$$

where $\boldsymbol{P}$ is the total linear momentum,

$$\boldsymbol{P} = \sum_i \boldsymbol{p}_i. \tag{5.328}$$

From (5.328) and (5.307) it follows that for any two components P_k and P_l of $\boldsymbol{P}$ we have $\{P_k, P_l\} = 0$.

We discussed in Chapter 2 the relationship between infinitesimal transformations and constants of motion. We shall consider this connexion again, but now from the point of view of the Hamiltonian equations of motion.

Let H be invariant under an infinitesimal transformation which is such that only one of the generalised coordinates (q_{k_0}, say) changes while the other coordinates and the momenta remain unchanged; this is, we have

$$\delta H = 0, \quad \text{if} \quad q_{k_0} \to q_{k_0} + \delta q_{k_0}, \tag{5.329}$$

or

$$\frac{\partial H}{\partial q_{k_0}} = 0. \tag{5.330}$$

From (5.330) and (5.315) it follows that

$$\{H, p_{k_0}\} = 0, \tag{5.331}$$

and thus that p_{k_0} is a constant of motion. (One could, of course, also combine (5.330) and (5.108) to shew that $\dot{p}_{k_0} = 0$, so that p_{k_0} is constant.)

It is often possible to find a class of transformations which include infinitesimal ones under which H is invariant; it is, however, not always easy to find the particular q_{k_0} which corresponds to these infinitesimal transformations, and it is therefore not always easy to see immediately which function of the coordinates and momenta will be a constant of motion.

The simplest cases, and the ones to which we shall restrict ourselves, are those of the translational or rotational invariance. In those two cases we shall be able to find the particular q_{k_0} corresponding to the infinitesimal transformations. In our discussion of translational or rotational invariance we shall assume that the Cartesian $\boldsymbol{x}_i$ $(i = 1, \ldots, N)$ form a suitable set of coordinates.

Let us first consider the case of translational invariance. This means that H is invariant under the transformation

$$\boldsymbol{x}_i \to \boldsymbol{x}_i + \boldsymbol{\varepsilon}, \qquad i = 1, \ldots, N. \tag{5.332}$$

The q_{k_0} for which we are looking are now, of course, the three components of the position vector of the centre of mass of the system so that the components of the total linear momentum vector $\boldsymbol{P}$ which is given by (5.328) are constants of motion. We can prove that as follows. On the one hand, we have

$$\{\boldsymbol{P}, H\} = \sum_i \{\boldsymbol{p}_i, H\} = -\sum_i \nabla_i H, \tag{5.333}$$

where we have used (5.315). On the other hand, we have for the change δH of the Hamiltonian under the transformation (5.332)

$$\delta H = \sum_i (\delta \boldsymbol{x}_i \cdot \nabla_i H) = (\boldsymbol{\varepsilon} \cdot \sum_i \nabla_i H), \tag{5.334}$$

and from the requirement that $\delta H = 0$ for any choice of $\boldsymbol{\varepsilon}$ it follows that $\sum_i \nabla_i H$ vanishes and thus from (5.333) that the three components of $\boldsymbol{P}$ are constants of motion.

We shall see presently that $\boldsymbol{P}$ can be said to generate a translation. It is of interest in this connexion to consider the change in a function f of the coordinates $\boldsymbol{x}_i$ under the transformation (5.332). We find

$$\delta f = \sum_i (\delta \boldsymbol{x}_i \cdot \nabla_i f) = (\boldsymbol{\varepsilon} \cdot \sum_i \nabla_i f) = (\boldsymbol{\varepsilon} \cdot \{f, \boldsymbol{P}\}). \tag{5.335}$$

The situation is very similar with regard to rotational invariance. We are now interested in a transformation

$$\boldsymbol{x}_i \to \boldsymbol{x}_i + [\boldsymbol{\varepsilon} \wedge \boldsymbol{x}_i], \qquad i = 1, \ldots, N; \qquad \boldsymbol{\varepsilon} = \boldsymbol{n}\delta\theta, \tag{5.336a}$$

which corresponds to a rotation around an axis parallel to the unit vector $\boldsymbol{n}$ over an angle $\delta\theta$. The transformation (5.336a) holds, of course, for all Cartesian vectors so that when we are considering the change in the Hamiltonian, or in another function of the $\boldsymbol{x}_i$ and the $\boldsymbol{p}_i$, we must bear in mind that the $\boldsymbol{p}_i$ are at the same time transformed, namely according to the equations

$$\boldsymbol{p}_i \to \boldsymbol{p}_i + [\boldsymbol{\varepsilon} \wedge \boldsymbol{p}_i], \qquad i = 1, \ldots, N. \tag{5.336b}$$

The q_{k_0} we are now trying to find are angles, and the corresponding constants of motion will be angular momenta. In the present case we have, first of all,

$$\begin{aligned}\{\boldsymbol{M}, H\} &= \{\sum_i [\boldsymbol{x}_i \wedge \boldsymbol{p}_i], H\} = -\sum_i [\boldsymbol{x}_i \wedge \nabla_i H] + \sum_i [\nabla_{pi} H \wedge \boldsymbol{p}_i] \\ &= -\sum_i [\boldsymbol{x}_i \wedge \nabla_i H],\end{aligned} \tag{5.337}$$

where ∇_{pi} is a symbolic vector with components $\partial/\partial p_{xi}$, $\partial/\partial p_{yi}$, and $\partial/\partial p_{zi}$, and where we have used the fact that for the case of Cartesian coordinates H contains the $\boldsymbol{p}_i$ in the combination $\boldsymbol{p}_i^2/2m_i$, so that $\nabla_{pi} H = \boldsymbol{p}_i/m_i$.

Instead of (5.334) we now have

$$\begin{aligned}\delta H &= \sum_i ([\boldsymbol{\varepsilon} \wedge \boldsymbol{x}_i] \cdot \nabla_i H) + \sum_i ([\boldsymbol{\varepsilon} \wedge \boldsymbol{p}_i] \cdot \nabla_{pi} H) \\ &= (\boldsymbol{\varepsilon} \cdot \sum_i [\boldsymbol{x}_i \wedge \nabla_i H]) + (\boldsymbol{\varepsilon} \cdot \sum_i [\boldsymbol{p}_i \wedge \nabla_{pi} H]) \\ &= (\boldsymbol{\varepsilon} \cdot \sum_i [\boldsymbol{x}_i \wedge \nabla_i H]),\end{aligned} \tag{5.338}$$

and the fact that δH is zero for any choice of $\boldsymbol{\varepsilon}$ leads to $\{\boldsymbol{M}, H\} = 0$, so that the three components of $\boldsymbol{M}$ are constants of motion, if H is invariant under rotations.

We shall discuss presently in what sense the angular momentum $\boldsymbol{M}$ can be said to generate rotations. It is of interest for our further discussion to find the change δf in a function $f(\boldsymbol{x}_i, \boldsymbol{p}_i)$ under the transformation (5.336). We find for δf the equation

$$\begin{aligned}\delta f &= \sum_i (\delta \boldsymbol{x}_i \cdot \nabla_i f) + \sum_i (\delta \boldsymbol{p}_i \cdot \nabla_{pi} f) \\ &= \sum_i ([\boldsymbol{\varepsilon} \wedge \boldsymbol{x}_i] \cdot \nabla_i f) + \sum_i ([\boldsymbol{\varepsilon} \wedge \boldsymbol{p}_i] \cdot \nabla_{pi} f) \\ &= (\boldsymbol{\varepsilon} \cdot \sum_i [\boldsymbol{x}_i \wedge \nabla_i f]) + (\boldsymbol{\varepsilon} \cdot \sum_i [\boldsymbol{p}_i \wedge \nabla_{pi} f]) \\ &= (\boldsymbol{\varepsilon} \cdot \{f, \boldsymbol{M}\}).\end{aligned} \tag{5.339}$$

Let us consider for a moment the situation where H is invariant both under translations and under rotations. Let us suppose that we had found the following constants of motion: M_x, M_y, and P_x. We could then have used our earlier result that the Poisson bracket of any two constants of motion is again a constant of motion and combined this with (5.325) to prove that M_z also should be a constant of motion, or with (5.327) to prove that P_z and then also P_y should be constants of motion. We see thus that, if two components of the angular momentum vector are constants of motion, the third component will also be a constant of motion. This is not true for the linear momentum vector, since $\{P_x, P_y\} = 0$. Once we have found P_x, P_y, P_z, M_x, M_y, and M_z, we have exhausted all possibilities, and no new constants of motion can be obtained from the Poisson brackets involving these six quantities.

From the fact that $\{M_x, M_y\} = M_z$ and the relation $\{p_k, p_l\} = 0$ it follows that two components of the angular momentum vector cannot simultaneously be canonical momenta. On the other hand $\{M^2, M_x\} = 0$ so that the absolute magnitude of the total angular momentum can be a canonical momentum at the same time as one of its components (compare the discussion of the central forces problem in § 2.3 and the well-known situation in quantum mechanics).

We saw in the previous section that the generating function

$$S = \sum_k \alpha_k q_k \tag{5.340}$$

would lead to the identical transformation, that is,

$$p_k = \frac{\partial S}{\partial q_k} = \alpha_k, \qquad \beta_k = \frac{\partial S}{\partial \alpha_k} = q_k. \tag{5.341}$$

Let us now consider an infinitesimal transformation generated by the function

$$S = \sum_k \alpha_k q_k + \varepsilon f(\alpha, q), \tag{5.342}$$

where ε is an infinitesimal quantity and f an arbitrary function of the α_k and q_k. The transformation generated by the function (5.342) is

$$p_k = \alpha_k + \varepsilon \frac{\partial f}{\partial q_k}, \qquad \beta_k = q_k + \varepsilon \frac{\partial f}{\partial \alpha_k}, \tag{5.343}$$

which, up to first order in ε, is equivalent to

$$\alpha_k = p_k - \varepsilon \frac{\partial f(p, q)}{\partial q_k}, \qquad \beta_k = q_k + \varepsilon \frac{\partial f(p, q)}{\partial p_k}. \tag{5.344}$$

We shall now consider a few special cases of infinitesimal transformations. The first one we wish to consider is the one where $\varepsilon = \delta t$ and $f = H(\alpha, q)$, where H is the Hamiltonian of the system. We then have from (5.344)

$$\beta_k - q_k = \delta q_k = \delta t \frac{\partial H}{\partial p_k}, \qquad \alpha_k - p_k = \delta p_k = -\delta t \frac{\partial H}{\partial q_k}, \tag{5.345}$$

or: *the Hamiltonian generates the motion of the system in phase space with time.*

As a second special case we choose $f = (\boldsymbol{n} \cdot \boldsymbol{P}(\boldsymbol{\alpha}))$ where $\boldsymbol{P}(\boldsymbol{p}_i)$ is the total linear momentum of the system which is given by (5.328), and where the $\boldsymbol{\alpha}$ are vectors which are related to the $\boldsymbol{p}_i$ in the same way as the α_k are related to the p_k. We can write the expression for f in the following way,

$$f = (\boldsymbol{n} \cdot \boldsymbol{P}(\boldsymbol{\alpha})) = (\boldsymbol{n} \cdot \sum_i \boldsymbol{\alpha}_i), \tag{5.346}$$

and we get from (5.344)

$$\boldsymbol{\alpha}_i = \boldsymbol{p}_i, \qquad \boldsymbol{\beta}_i = \boldsymbol{x}_i + \boldsymbol{\varepsilon}, \tag{5.347}$$

where $\boldsymbol{\varepsilon} = \varepsilon \boldsymbol{n}$, and where the $\boldsymbol{\beta}_i$ are the vectors which are canonically conjugate to the $\boldsymbol{\alpha}_i$. Equations (5.347) are the same as (5.332) which described a translation.

The last example is one where we put

$$f = (\boldsymbol{n} \cdot \boldsymbol{M}(\boldsymbol{\alpha}, \boldsymbol{x})) = (\boldsymbol{n} \cdot \sum_i [\boldsymbol{x}_i \wedge \boldsymbol{\alpha}_i]), \tag{5.348}$$

where $\boldsymbol{M}(\boldsymbol{p}, \boldsymbol{x})$ is the total angular momentum given by (5.324). From (5.344) we now get

$$\boldsymbol{\alpha}_i = \boldsymbol{p}_i + [\boldsymbol{\varepsilon} \wedge \boldsymbol{p}_i], \qquad \boldsymbol{\beta}_i = \boldsymbol{x}_i + [\boldsymbol{\varepsilon} \wedge \boldsymbol{x}_i], \tag{5.349}$$

where again $\boldsymbol{\varepsilon} = \varepsilon \boldsymbol{n}$. Equations (5.349) are the same as (5.336) and correspond thus to a rotation.

The present discussion shews what was meant by the statement that the total linear and angular momenta generate respectively translations and rotations.

The change in any function $F(p, q)$ of the p_k and q_k under the transformation generated by the function S of (5.342) is given by the equation

$$\delta F = \sum_k \left[\frac{\partial F}{\partial q_k}\,\delta q_k + \frac{\partial F}{\partial p_k}\,\delta p_k\right]$$

$$= \varepsilon \sum_k \left[\frac{\partial F}{\partial q_k}\frac{\partial f}{\partial p_k} - \frac{\partial F}{\partial p_k}\frac{\partial f}{\partial q_k}\right] = \varepsilon\{F, f\}. \tag{5.350}$$

We can compare this expression with (5.335) and (5.339) which are obtained by substituting respectively expressions (5.346) and (5.348) for f into (5.350), and using the fact that $\boldsymbol{\varepsilon} = \varepsilon \boldsymbol{n}$.

An interesting special case arises when we substitute into (5.350) the Hamiltonian H for F. We then have

$$\delta H = \varepsilon\{H, f\}, \tag{5.351}$$

and we see that any constant of motion will generate an infinitesimal transformation which leaves H invariant, because if f is an integral of motion, $\{H, f\} = 0$. This is the inverse of the theorem we met with a little earlier that one can always find an integral of motion, if we know infinitesimal transformations which leave H invariant.

To conclude this section we shall consider the case of a point particle in an electromagnetic field, and use the equations of motion in the form following from (5.108) and (5.315),

$$\dot{q}_k = \{q_k, H\}, \qquad \dot{p}_k = \{p_k, H\}, \tag{5.352}$$

which are special cases of (5.303). The Hamiltonian in this case follows from the Lagrangian [see (2.507)]

$$L = \tfrac{1}{2}m\dot{\boldsymbol{x}}^2 - e\phi + e(\boldsymbol{A}\cdot\dot{\boldsymbol{x}}), \tag{5.353}$$

and we get

$$\boldsymbol{p} = m\dot{\boldsymbol{x}} + e\boldsymbol{A}, \tag{5.354}$$

and

$$H = (\boldsymbol{p}\cdot\dot{\boldsymbol{x}}) - L = (\boldsymbol{p} - e\boldsymbol{A})^2/2m + e\phi. \tag{5.355}$$

The equations of motion are now

$$\dot{\boldsymbol{x}} = \{\boldsymbol{x}, H\}, \qquad \dot{\boldsymbol{p}} = \{\boldsymbol{p}, H\}, \tag{5.356}$$

or

$$\dot{\boldsymbol{x}} = (\boldsymbol{p} - e\boldsymbol{A})/m, \tag{5.357}$$

which is the same as (5.354), and

$$\dot{\boldsymbol{p}} = -e\nabla\phi + (e/m)[(\boldsymbol{p} - e\boldsymbol{A}) \wedge [\nabla \wedge \boldsymbol{A}]], \tag{5.358}$$

which is equivalent to (2.509) and which thus can be reduced to the equation

$$m\ddot{x} = e\{E+[x \wedge B]\}, \qquad E = -\nabla\phi - \frac{\partial A}{\partial t}, \qquad B = [\nabla \wedge A]. \qquad (5.359)$$

4. VARIATIONAL PRINCIPLES; TIME AND ENERGY AS CANONICALLY CONJUGATE VARIABLES

In Chapter 2 we proved that d'Alembert's principle could be expressed in the form (2.229) which, if we use (2.232) is equivalent to

$$\int_{t_1}^{t_2} \delta(T-U)\mathrm{d}t = \left[\sum_i m_i(\dot{x}_i \cdot \delta x_i)\right]_{t_1}^{t_2}, \qquad (5.401)$$

where the δx_i are any variation of the x_i which are compatible with the kinematic relations. For our present discussion it is also important to emphasise that in deriving (5.401) we were considering variations where the time was *not* varied, that is, we compared points on the original and the new orbits at the same time (points A_n and B_n in fig. 25).

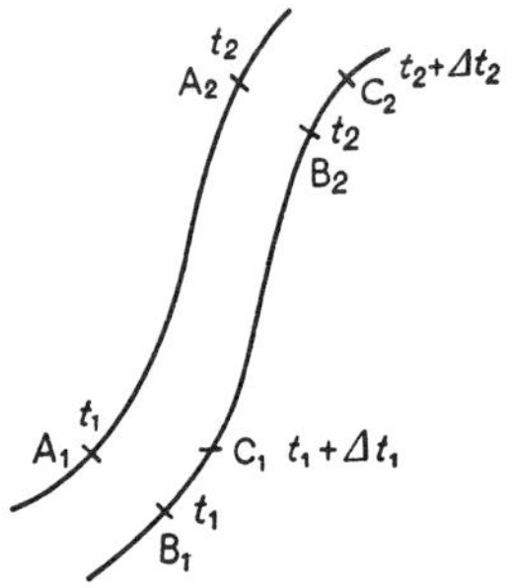

Fig. 25. Variation of orbits. $A_1, A_2, \ldots$ are points on the original orbit which are passed at times $t_1, t_2, \ldots$; $B_1, B_2, \ldots$ are points on the new orbit which are passed at the same times $t_1, t_2, \ldots$; $C_1, C_2, \ldots$ are points on the new orbit which are passed at the varied times $t_1+\Delta t_1, t_2+\Delta t_2, \ldots$.

We saw in Chapter 2 that, if we compare two possible orbits such that $\delta x_i = 0$ both at t_1 and at t_2, equation (5.401) is equivalent to

$$\int_{t_1}^{t_2} \delta L \mathrm{d}t = 0, \qquad (5.402)$$

from which the Lagrangian equations of motion (2.308) follow straightforwardly.

We shall now consider a different variation of the orbit, namely one where we compare the $\boldsymbol{x}_i$ at time t with $\boldsymbol{x}_i + \Delta\boldsymbol{x}_i$ at time $t+\Delta t$ (points A_n and C_n in fig. 25). To each point A_n of the original orbit there thus corresponds a point B_n at the same time on the new orbit as well as a point C_n at the varied time on the new orbit, and we write

$$\boldsymbol{x}_{iB} = \boldsymbol{x}_{iA} + \delta\boldsymbol{x}_i, \qquad \boldsymbol{x}_{iC} = \boldsymbol{x}_{iA} + \Delta\boldsymbol{x}_i, \tag{5.403}$$

where the $\delta\boldsymbol{x}_i$ and the $\Delta\boldsymbol{x}_i$ are related through the equation

$$\Delta\boldsymbol{x}_i = \delta\boldsymbol{x}_i + \dot{\boldsymbol{x}}_i\Delta t. \tag{5.404}$$

Equation (5.402) was derived for the A → B-variation. If we wish to get a variational principle for the A → C-variation, we must take into account that now the limits of our integral are also varying.

We first of all define the *action integral I* by the equation

$$I = \int_{t_1}^{t_2} 2T\,\mathrm{d}t. \tag{5.405}$$

Its variation is given by the equation

$$\delta I = \delta\int_{t_1}^{t_2} 2T\,\mathrm{d}t = \int_{t_1}^{t_2} \delta(2T)\mathrm{d}t + 2T\delta t\Big|_{t_1}^{t_2}$$

$$= \int_{t_1}^{t_2} \delta(T+U)\mathrm{d}t + \int_{t_1}^{t_2} \delta(T-U)\mathrm{d}t + 2T\Delta t\Big|_{t_1}^{t_2},$$

or,

$$\delta I = \int_{t_1}^{t_2} \delta E\,\mathrm{d}t + \sum_i m_i(\dot{\boldsymbol{x}}_i\cdot\Delta\boldsymbol{x}_i)\Big|_{t_1}^{t_2}, \tag{5.406}$$

where we have used (5.401) and (5.404), the fact that $\delta t \equiv \Delta t$, and the basic equation for the variation of an integral

$$\delta\int_p^q f(x)\mathrm{d}x = \int_p^q \delta f\,\mathrm{d}x + f(q)\delta q - f(p)\delta p. \tag{5.407}$$

The reason why we can use (5.401) with the $\delta\boldsymbol{x}_i$ (and not the $\Delta\boldsymbol{x}_i$) is that the variation of the time is taken into account in the term $2T\Delta t$ at the two end points; put differently: we first compare those parts of the original and the new orbit where t goes through the same interval and we are then left with the contribution from the segments at the two ends of the orbits.

The first conclusion we can draw from (5.406) is one for the case of two periodic orbits, which are both possible orbits which means that the total energy E is a constant of motion and therefore does not vary over the orbit. If τ is the period of the orbit, we have

$$\begin{aligned}\delta I &= \int_{t_1}^{t_1+\tau} \delta E\,\mathrm{d}t + \sum_i m_i(\dot{\boldsymbol{x}}_i \cdot \Delta \boldsymbol{x}_i)\Big|_{t_1}^{t_1+\tau} \\ &= \delta E \int_{t_1}^{t_1+\tau} \mathrm{d}t = \tau \delta E, \end{aligned} \tag{5.408}$$

or,

$$\tau = \partial I / \partial E, \tag{5.409}$$

where now I is the action integral extended over one complete orbit. The integrated term vanished, because we were dealing with two periodic orbits so that the terms at the upper and the lower limit cancelled.

The relation (5.409) is of interest in the one-dimensional case, where we can write $p\dot{q}$ for $2T$ [compare (2.302) and (2.310)] so that we have for I

$$I = \int_{t_1}^{t_1+\tau} p\dot{q}\,\mathrm{d}t = \int_{t_1}^{t_1+\tau} p\,\mathrm{d}q = \oint p\,\mathrm{d}q, \tag{5.410}$$

where $\oint$ indicates an integration over one complete period. According to the old quantum theory the integral on the right hand side of (5.410) is quantised and put equal to nh (h: Planck's constant), or,

$$I = nh, \tag{5.411}$$

and if we apply (5.408) to finite charges in I and E, ΔI and ΔE, we find

$$\Delta I = h, \quad \text{and} \quad E = h\omega, \tag{5.412}$$

where

$$\omega = 1/\tau. \tag{5.413}$$

We may refer here also to the discussion of action and angle variables in § 6.2.

Another application of (5.406) occurs when we compare two orbits with the same energy ($\delta E = 0$) and with the same starting point and final point [$\Delta \boldsymbol{x}_i = 0$, $t = t_1$ or $t = t_2$; note that these two end points are not reached at the same time in the original and the new orbit: compare the discussion of (5.406)]. In that case we get the *principle of least action*:

$$\delta I = 0, \text{ if } \delta E = 0 \text{ and } \Delta \boldsymbol{x}_i = 0 \text{ for } t = t_1 \text{ and } t = t_2. \tag{5.414}$$

We see here that if we consider two slightly different orbits with the same end points there are two possibilities:

(i) if the time is not varied, $\delta \int L \mathrm{d}t = 0,$

(ii) if the energy is not varied, $\delta \int 2T \mathrm{d}t = 0.$

We have seen how the Lagrangian equations of motion follow from the first variational principle, and we shall see in a moment the connexion between the second variational principle and the canonical equations of motion.

To see this connexion we shall prove that the canonical equations of motion are equivalent to the variational principle

$$\delta \int_1^2 \sum_k p_k \mathrm{d}q_k = 0 \quad (\delta q_k = 0, \quad \delta p_k = 0 \text{ at end points}), \tag{5.415}$$

with the condition

$$\delta H = 0 \text{ at each point of the orbit.} \tag{5.416}$$

The difference with the principle of least action is now that we consider variations where the $2s$ variables p_k, q_k are all independent.

Let there be a parameter u which determines the position along the orbit, so that the δq_k and δp_k are functions of u. Equation (5.415) can be written in the form

$$\begin{aligned}
\delta \int \sum_k p_k \mathrm{d}q_k &= \int (\sum_k \delta p_k \mathrm{d}q_k + \sum_k p_k \delta \mathrm{d}q_k) \\
&= \int \sum_k (\delta p_k \mathrm{d}q_k - \delta q_k \mathrm{d}p_k) + p_k \delta q_k \Big|_1^2 \\
&= \int \sum_k \left(\delta p_k \frac{\mathrm{d}q_k}{\mathrm{d}u} - \delta q_k \frac{\mathrm{d}p_k}{\mathrm{d}u} \right) \mathrm{d}u.
\end{aligned} \tag{5.417}$$

Equation (5.416) can be written as

$$\sum_k \left(\frac{\partial H}{\partial p_k} \delta p_k + \frac{\partial H}{\partial q_k} \delta q_k \right) = 0, \tag{5.418}$$

and this equation must hold at any point of the orbit, that is, for all values for u. Using the method of the Lagrangian multipliers we get

$$\int \sum_k \left[\delta p_k \left(\frac{\mathrm{d}q_k}{\mathrm{d}u} - \lambda(u)\frac{\partial H}{\partial p_k}\right) - \delta q_k \left(\frac{\mathrm{d}p_k}{\mathrm{d}u} + \lambda(u)\frac{\partial H}{\partial q_k}\right)\right] \mathrm{d}u = 0, \tag{5.419}$$

or

$$\frac{\mathrm{d}q_k}{\mathrm{d}u} = \lambda(u)\frac{\partial H}{\partial p_k}, \qquad \frac{\mathrm{d}p_k}{\mathrm{d}u} = -\lambda(u)\frac{\partial H}{\partial q_k}. \tag{5.420}$$

These equations give us the direction of the orbit in the $2s$-dimensional phase space. If we introduce a variable t by

$$\mathrm{d}t = \lambda(u)\mathrm{d}u, \quad \text{or,} \quad t = \int^u \lambda(u)\mathrm{d}u, \tag{5.421}$$

equations (5.420) are reduced to the canonical equations of motion

$$\frac{\mathrm{d}q_k}{\mathrm{d}t} = \frac{\partial H}{\partial p_k}, \qquad \frac{\mathrm{d}p_k}{\mathrm{d}t} = -\frac{\partial H}{\partial q_k}. \tag{5.422}$$

The relation with the principle of least action becomes even more apparent if we write [compare (5.109)]

$$\sum_k p_k \mathrm{d}q_k = \sum_k p_k \dot{q}_k \mathrm{d}t = 2T\mathrm{d}t. \tag{5.423}$$

We shall conclude this chapter by shewing how one can introduce the time and the (negative of the) energy as canonically conjugate variables. We shall for this discussion drop the restriction that the Hamiltonian does not contain the time t explicitly. One can easily convince oneself that the canonical equations of motion (5.108) are still valid, but instead of (5.111) we now have

$$\frac{\mathrm{d}H}{\mathrm{d}t} = \sum_k \left(\frac{\partial H}{\partial p_k}\dot{p}_k + \frac{\partial H}{\partial q_k}\dot{q}_k\right) + \frac{\partial H}{\partial t} = \frac{\partial H}{\partial t}. \tag{5.424}$$

The most satisfactory method to introduce the time as one of the q's is to use Hamilton's variational principle in the form (5.215). This variational principle was

$$\delta\int_1^2 L\mathrm{d}t = \delta\int_1^2 [\sum_k p_k\dot{q}_k - H(p_k, q_k, t)]\mathrm{d}t = 0, \tag{5.425}$$

where now the p_k and q_k from a set of $2s$ independent variables. We now introduce a function p_0 and also change the variable of integration from t to u, where u may be any function of the time which determines the position of the representative point along its orbit in phase space. We then get the variational principle

$$\delta \int_1^2 \left(\sum_{k=1}^{s} p_k q_k' + p_0 q_0' \right) du, \tag{5.426}$$

with the auxiliary condition

$$p_0 = -H(p_k, q_k, q_0). \tag{5.427}$$

The primes in (5.426) indicate differentiation with respect to u and we have replaced the explicitly appearing time coordinate by q_0:

$$q_k' = \frac{dq_k}{du}, \qquad q_0 = t, \qquad q_0' = \frac{dt}{du}. \tag{5.428}$$

Equation (5.427) can be written in the form

$$\mathscr{H} \equiv p_0 + H = 0, \quad \text{or,} \quad \delta\mathscr{H} = 0. \tag{5.429}$$

From (5.426) and (5.429) we get the equations

$$\frac{dp_k}{du} = -\lambda \frac{\partial\mathscr{H}}{\partial q_k}, \qquad \frac{dq_k}{du} = \lambda \frac{\partial\mathscr{H}}{\partial p_k}, \tag{5.430}$$

just as (5.420) followed from (5.415) and (5.416). Introducing the time through (5.421) we get from (5.430)

$$\frac{dp_k}{dt} = -\frac{\partial\mathscr{H}}{\partial q_k} = -\frac{\partial H}{\partial q_k}, \qquad \frac{dq_k}{dt} = \frac{\partial\mathscr{H}}{\partial p_k} = \frac{\partial H}{\partial p_k}, \qquad k = 1, \ldots, s, \tag{5.431}$$

which are the canonical equations of motion, and

$$\frac{dp_0}{dt} = -\frac{\partial\mathscr{H}}{\partial q_0} = -\frac{\partial H}{\partial q_0}, \qquad \frac{dq_0}{dt} = \frac{\partial\mathscr{H}}{\partial p_0} = 1. \tag{5.432}$$

From the second of equations (5.432) we see that, indeed, q_0 can be identified with the time, while from (5.431) and (5.432) it follows that

$$\frac{d\mathscr{H}}{dt} = \sum_{k=0}^{s} \left(\frac{\partial\mathscr{H}}{\partial q_k} \dot{q}_k + \frac{\partial\mathscr{H}}{\partial p_k} \dot{p}_k \right) = 0, \tag{5.433}$$

since there is now no explicitly-appearing t (its place being taken by q_0). From (5.433) it follows that

$$\mathscr{H} = \text{constant}, \tag{5.434}$$

and we can without loss of generality put this constant equal to zero so that $p_0 = -H$. As p_0, q_0 appear in the same way as the other p_k, q_k we see that, indeed $-H$ $(= p_0)$ is canonically conjugate to t $(= q_0)$. In non-relativistic mechanics this result is not particularly important, but it has great importance in relativistic mechanics, since the time there occurs on equal footing with the spatial coordinates.

We must, in conclusion, draw attention to the fact that instead of a Hamiltonian function H we now have a Hamiltonian equation $\mathscr{H} = 0$. The left hand side of this Hamiltonian equation occurs in the equations of motion (5.431) and (5.432). We must warn that sometimes one cannot use immediately the left hand side of a given Hamiltonian equation to obtain the canonical equations of motion. In order that the function $\mathscr{H}$ occurring in these equations is really the function on the left hand side of the Hamiltonian equation it is necessary that p_0 occurs linearly in $\mathscr{H}$ and with coefficient unity, because otherwise the second of equations (5.432) will not hold, and q_0 will not be the time. To illustrate this point we consider the Hamiltonian equation for a point particle in an electromagnetic field — the relativistic relation between the four components of the energy-momentum four-vector of a particle in an electromagnetic field —,

$$0 = \mathscr{H}' = (p_0 + e\phi)^2 - c^2[(\boldsymbol{p} - e\boldsymbol{A})^2 + m^2c^2], \tag{5.435}$$

where ϕ and $\boldsymbol{A}$ are again the scalar and vector potentials of the electromagnetic field. Forming a function $\mathscr{H}$ which has the dimensions of an energy and which contains p_0 linearly and with coefficient 1 we get a new Hamiltonian equation:

$$0 = \mathscr{H} = p_0 + e\phi - c[(\boldsymbol{p} - e\boldsymbol{A})^2 + m^2c^2]^{\frac{1}{2}}. \tag{5.436}$$

We leave it to the reader to prove that (5.431) and (5.432) with $\mathscr{H}$ given by (5.436) lead to the relativistic equations of motion [compare the derivation of (2.509)]

$$\frac{\mathrm{d}}{\mathrm{d}t} \frac{m\dot{\boldsymbol{x}}}{\sqrt{1 - (v^2/c^2)}} = e\{\boldsymbol{E} + [\dot{\boldsymbol{x}} \wedge \boldsymbol{B}]\}, \qquad v^2 = (\dot{\boldsymbol{x}} \cdot \dot{\boldsymbol{x}}), \tag{5.437}$$

where $\boldsymbol{E}$ and $\boldsymbol{B}$ are again the electrical field strength and the magnetic induction.

CHAPTER 6

HAMILTON-JACOBI THEORY

In this chapter we introduce the Hamilton-Jacobi function which is a solution of the Hamilton-Jacobi partial differential equation and which leads to a Hamiltonian containing only one set of canonical variables. The Hamilton-Jacobi equation is solved for some simple cases, including that of the Kepler problem. In the second section we discuss the so-called action and angle variables. Their importance derives from the fact that the action variables are adiabatic invariants. These played a large role in the old quantum theory and are also important in the theory of particle accelerators. They are briefly discussed in the last section.

1. THE HAMILTON-JACOBI EQUATION

In the previous chapter we discussed various transformations which left the canonical equations (5.108) invariant in form and which were obtained using generating functions. We mentioned there that we would especially consider transformations of the type (5.220b). The purpose of these transformations is to simplify the equations of motion. This is achieved, if the transformation leads to a transformed Hamiltonian which depends on only one set of the canonical variables (say, the α_k) while not containing the other set (β_k). If we have such a Hamiltonian $\overline{H}(\alpha_k)$ the equations of motion become

$$\dot{\alpha}_k = -\frac{\partial \overline{H}}{\partial \beta_k} = 0, \qquad \alpha_k = \text{constant}, \tag{6.101}$$

$$\dot{\beta}_k = \frac{\partial \overline{H}}{\partial \alpha_k} = \text{constant} = \gamma_k, \qquad \beta_k = \gamma_k t + \delta_k, \tag{6.102}$$

where the second set of equations follows from the fact that the α_k are constants according to (6.101) and that $\overline{H}$ is a function of the α_k only. The γ_k

are thus known functions of the α_k and the $2s$ integration constants are the α_k and the δ_k.

Our task is thus to find a generating function $S(\alpha_k, q_k)$ such that the transformation

$$p_k = \frac{\partial S}{\partial q_k}, \qquad \beta_k = \frac{\partial S}{\partial \alpha_k} \tag{6.103}$$

changes the Hamiltonian $H(p_k, q_k)$ into $\overline{H}(\alpha_k)$ which is independent of the β_k. This means that S must satisfy the following partial differential equation:

$$H\left(\frac{\partial S}{\partial q_k}, q_k\right) = \overline{H}(\alpha_k) = E(\alpha_k). \tag{6.104}$$

Equation (6.104) is the *Hamilton-Jacobi equation.* To emphasise that the right hand side of (6.104) is the total energy of the system we have called it E. It is often difficult to solve the Hamilton-Jacobi equation, but once we have found S the solution of the transformed equations of motion is trivial and is given by (6.101) and (6.102). We are still left then with the problem to find the original p_k, q_k as functions of the time and the transformation from the α_k, β_k to the p_k, q_k is often a complicated one (see, for instance, the discussion of the Kepler problem later in this section).

To see the physical meaning of S we evaluate $\mathrm{d}S/\mathrm{d}t$,

$$\frac{\mathrm{d}S}{\mathrm{d}t} = \sum_k \frac{\partial S}{\partial q_k}\dot{q}_k + \sum_k \frac{\partial S}{\partial \alpha_k}\dot{\alpha}_k = \sum_k p_k\dot{q}_k + \sum_k \beta_k\dot{\alpha}_k = \sum_k p_k\dot{q}_k, \tag{6.105}$$

where we have used (6.103) and (6.101). From (6.105) it follows that

$$S = \int^t \sum_k p_k\dot{q}_k \mathrm{d}t = I, \tag{6.106}$$

where I is the action integral defined by (5.405) [compare (5.109)].

We must emphasise here that our treatment only applies to the (most often encountered) case where the Hamiltonian does not contain the time explicitly. If the Hamiltonian does contain the time explicitly, we can introduce the time as q_0, and use the Hamiltonian equation

$$H(p_k, q_k, q_0) + p_0 = 0, \tag{6.107}$$

which with a generating function $\bar{S}(q_k, \alpha_k; q_0, \alpha_0)$ leads to the equation

$$H\left(\frac{\partial \bar{S}}{\partial q_k}, q_k, q_0\right) + \frac{\partial \bar{S}}{\partial q_0} = 0, \tag{6.108}$$

or,

$$H\left(\frac{\partial \bar{S}}{\partial q_k}, q_k, t\right) + \frac{\partial \bar{S}}{\partial t} = 0. \tag{6.109}$$

It should be noted that (6.109) is often referred to as the Hamilton-Jacobi equation. To find the physical meaning of $\bar{S}$ we now evaluate $\mathrm{d}\bar{S}/\mathrm{d}t$,

$$\frac{\mathrm{d}\bar{S}}{\mathrm{d}t} = \sum_{k=1}^{s} \frac{\partial \bar{S}}{\partial q_k}\dot{q}_k + \sum_{k=1}^{s} \frac{\partial \bar{S}}{\partial \alpha_k}\dot{\alpha}_k + \frac{\partial \bar{S}}{\partial q_0}\dot{q}_0 + \frac{\partial \bar{S}}{\partial \alpha_0}\dot{\alpha}_0 = \sum_{k=1}^{s} p_k\dot{q}_k + p_0, \tag{6.110}$$

or,

$$\begin{aligned}\bar{S} &= \int^t (\sum_k p_k\dot{q}_k + p_0)\mathrm{d}t \\ &= \int^t (\sum_k p_k\dot{q}_k - H)\mathrm{d}t = \int^t L\mathrm{d}t.\end{aligned} \tag{6.111}$$

If H does not contain the time explicitly, H is a constant which we may denote by E, and (6.111) leads to

$$\bar{S} = I - Et = S - Et, \tag{6.112}$$

while (6.109) reduces to (6.104).

At this point we remind the reader of the relation between the Hamilton-Jacobi function and the Schrödinger wave function. To see this relation we shall only consider the one-dimensional case where the Hamiltonian is given by the expression

$$H = (p^2/2m) + U(q),$$

which leads to the Schrödinger equation

$$-\frac{\hbar^2}{2m}\frac{\partial^2 \psi}{\partial q^2} + U(q)\psi = E\psi,$$

where ψ is the wave-function.

If we substitute in this time-independent Schrödinger equation

$$\psi = \mathrm{e}^{\mathrm{i}S/\hbar}, \tag{6.113}$$

we get the equation

$$\frac{1}{2m}\left(\frac{\partial S}{\partial q}\right)^2 - \frac{\mathrm{i}\hbar}{2m}\frac{\partial^2 S}{\partial q^2} + U(q) = E,$$

which in the limit as $\hbar \to 0$ reduces to the one-dimensional Hamilton-Jacobi equation [compare (6.114)].

If we substitute into (6.113) instead of S the function $\bar{S}$ of (6.112) we get instead of the time-independent wave function the time-dependent one.

We finally remind the reader that (6.113) is also the starting point for the WKB-approximation which is often called the semi-classical approximation for reasons which should be more obvious from the discussion just given.

We shall now consider three simple cases to illustrate how the Hamilton-Jacobi theory can be applied. These cases are the harmonic oscillator (in one dimension and in three dimensions), a point particle in a uniform gravitational field, and the Kepler problem.

We first of all consider the one-dimensional harmonic oscillator. In the general one-dimensional case the Hamiltonian is

$$H(p, q) = (p^2/2m)+U(q),$$

which leads to the Hamilton-Jacobi equation

$$(\partial S/\partial q)^2+2mU(q) = 2mE(\alpha) = 2m\alpha, \tag{6.114}$$

where we have chosen

$$E(\alpha) = \alpha. \tag{6.115}$$

In this case the Hamilton-Jacobi equation is particularly simple because there is only one q and the Hamilton-Jacobi equation gives us immediately $\partial S/\partial q$ as a function of q and of the parameter α. Quadrature leads directly to the result

$$S = \int^q [2m(\alpha-U)]^{\frac{1}{2}}\mathrm{d}q. \tag{6.116}$$

Using equations (6.103) we get

$$\beta = \partial S/\partial\alpha = \int_{q_0}^{q} [m/2(\alpha-U)]^{\frac{1}{2}}\mathrm{d}q, \tag{6.117}$$

$$p = \partial S/\partial q = [2m(\alpha-U)]^{\frac{1}{2}}. \tag{6.118}$$

It further follows from (6.101) and (6.102) that

$$\alpha = \text{constant}, \tag{6.119}$$

$$\beta = \partial\overline{H}/\partial\alpha = 1, \qquad \beta = t-t_0. \tag{6.120}$$

From (6.117) and (6.120) we obtain q as a function of the time. Equations (6.117) to (6.120) are the general solutions for the one-dimensional case. One sees immediately that (6.117) is the same as (1.120) while (6.118) is the energy equation.

In the case of the harmonic oscillator we have

$$U = \tfrac{1}{2}aq^2, \tag{6.121}$$

and we get from (6.117), (6.120), and (6.121)

$$q-q_0 = [2\alpha/a]^{\frac{1}{2}} \sin [(a/m)^{\frac{1}{2}}(t-t_0)], \tag{6.122}$$

which is the well-known solution of the harmonic oscillator problem.

Let us now consider the case of the three-dimensional harmonic oscillator; this corresponds to the Hamiltonian

$$H = [(p_1^2+p_2^2+p_3^2)/2m]+\tfrac{1}{2}a_1q_1^2+\tfrac{1}{2}a_2q_2^2+\tfrac{1}{2}a_3q_3^2, \tag{6.123}$$

leading to the following Hamilton-Jacobi equation

$$\left(\frac{\partial S}{\partial q_1}\right)^2+\left(\frac{\partial S}{\partial q_2}\right)^2+\left(\frac{\partial S}{\partial q_3}\right)^2+ma_1q_1^2+ma_2q_2^2+ma_3q_3^2 = 2mE(\alpha_k). \tag{6.124}$$

This equation can be solved by putting

$$S(q_1, q_2, q_3, \alpha_1, \alpha_2, \alpha_3) = S_1(q_1, \alpha_1)+S_2(q_2, \alpha_2)+S_3(q_3, \alpha_3). \tag{6.125}$$

This method of solving the Hamilton-Jacobi equation is the method of the *separation of variable.* In the general case where one can separate the variables the S_k will contain only one q_k each, but they may each contain all the α_k; in the present simple case each S_k contains one α_k only, but we shall encounter the more general case later in this section.

We draw attention to the fact that the separation of variables in the Hamilton-Jacobi case occurs by writing S as a sum of terms containing only one q_k, while in the case of the Schrödinger equation the separation occurs by writing the wave function as a product of such terms. This is, of course, a consequence of the relation (6.113) between the wave function and the Hamilton-Jacobi function. We also mention here that if for a physical system the Hamilton-Jacobi equation can be solved in a particular set of coordinates, the same set of coordinates can be used to separate the Schrödinger equation.

Substituting expression (6.125) into equation (6.124) we get three equations for the three S_k,

$$\left(\frac{\partial S_k}{\partial q_k}\right)^2+ma_kq_k^2 = 2m\alpha_k, \tag{6.126}$$

and $E(\alpha_k) = \alpha_1+\alpha_2+\alpha_3$: for each of the three degrees of freedom we are led to equations completely analogous to (6.116) to (6.122).

The second case is that of a point particle in a uniform gravitational field. The Hamiltonian is now

$$H = (p_x^2+p_y^2+p_z^2)/2m+mgz, \tag{6.127}$$

and the Hamilton-Jacobi equation is

$$E = \frac{1}{2m}\left[\left(\frac{\partial S}{\partial x}\right)^2 + \left(\frac{\partial S}{\partial y}\right)^2 + \left(\frac{\partial S}{\partial z}\right)^2\right] + mgz, \tag{6.128}$$

which, if we put

$$S = S_1(x) + S_2(y) + S_3(z), \tag{6.129}$$

and separate the variables, leads to

$$\frac{\partial S_1}{\partial x} = \alpha_1, \qquad \frac{\partial S_2}{\partial y} = \alpha_2, \tag{6.130}$$

$$[2m(\alpha_3 - mgz) - \alpha_1^2 - \alpha_2^2]^{\frac{1}{2}} = \frac{\partial S_3}{\partial z}, \tag{6.131}$$

or,

$$S = \int_{x_0}^{x} \alpha_1 \,\mathrm{d}x + \int_{y_0}^{y} \alpha_2 \,\mathrm{d}y + \int_{z_0}^{z} [2m(\alpha_3 - mgz) - \alpha_1^2 - \alpha_2^2]^{\frac{1}{2}} \mathrm{d}z. \tag{6.132}$$

As we have put $E = \alpha_3$, we get from (6.102) and (6.103)

$$\beta_3 = t - t_0 = \int_{z_0}^{z} m \,\mathrm{d}z / [2m(\alpha_3 - mgz) - \alpha_1^2 - \alpha_2^2]^{\frac{1}{2}}, \tag{6.133}$$

while

$$\begin{aligned} \beta_1 &= \text{constant} = x - x_0 - \int_{z_0}^{z} \alpha_1 \,\mathrm{d}z / [2m(\alpha_3 - mgz) - \alpha_1^2 - \alpha_2^2]^{\frac{1}{2}}, \\ \beta_2 &= \text{constant} = y - y_0 - \int_{z_0}^{z} \alpha_2 \,\mathrm{d}z / [2m(\alpha_3 - mgz) - \alpha_1^2 - \alpha_2^2]^{\frac{1}{2}}. \end{aligned} \tag{6.134}$$

If x_0, y_0, z_0 is the position of the particle at $t = t_0$, $\beta_1 = \beta_2 = 0$, and we get from (6.133) and (6.134)

$$\begin{aligned} x - x_0 &= (\alpha_1/m)(t - t_0), \\ y - y_0 &= (\alpha_2/m)(t - t_0), \\ z - z_0 &= [2m(\alpha_3 - mgz_0) - \alpha_1^2 - \alpha_2^2]^{\frac{1}{2}}(t - t_0)/m - \tfrac{1}{2}g(t - t_0)^2, \end{aligned} \tag{6.135}$$

which corresponds to a parabola, as should have been expected. The linear terms on the right hand side of equations (6.135) correspond to the uniform motion which the particle would shew, if there were no acceleration.

This last example is one where all but one of the coordinates were ignorable. In that case one can always solve the Hamilton-Jacobi equation by separation of variables, putting the momenta corresponding to the $s-1$ ignorable coordinates equal to $\alpha_1, \ldots, \alpha_{s-1}$; the last part of the Hamilton-Jacobi function can then be obtained by a simple quadrature.

Our last example is the Kepler problem. In spherical polars the Hamiltonian for the problem is [compare (2.313) to (2.317)]

$$H = \frac{p_r^2}{2m} + \frac{p_\theta^2}{2mr^2} + \frac{p_\varphi^2}{2mr^2 \sin^2\theta} - \frac{Ze^2}{4\pi\varepsilon_0 r}, \tag{6.136}$$

where to fix our ideas we have chosen the problem of a hydrogen-like atom, that is, an electron of charge $-e$ moving in the field of a nucleus with charge Ze. The mass m entering into (6.136) is really the reduced mass (see the discussion in § 1.2).

Introducing the Hamilton-Jacobi function $S(r, \theta, \varphi)$ and using the method of the separation of variables, we put

$$S(r, \theta, \varphi) = S_1(r) + S_2(\theta) + S_3(\varphi). \tag{6.137}$$

Putting once again the total energy equal to one of the α_k, α_1' say, we have the following equation

$$\alpha_1' = \frac{1}{2m}\left[\left(\frac{\partial S_1}{\partial r}\right)^2 + \frac{1}{r^2}\left(\frac{\partial S_2}{\partial \theta}\right)^2 + \frac{1}{r^2 \sin^2\theta}\left(\frac{\partial S_3}{\partial \varphi}\right)^2\right] - \frac{Ze^2}{4\pi\varepsilon_0 r}, \tag{6.138}$$

where we have used (6.103). We have put a prime on α_1' because it will be more convenient later to use a function of α_1' as α_1 [see (6.143)].

Separating this equation we are led to the equations

$$\frac{\partial S_3}{\partial \varphi} = \alpha_3 = \text{constant} = p_\varphi, \qquad S_3 = \varphi\alpha_3; \tag{6.139}$$

$$\left(\frac{\partial S_2}{\partial \theta}\right)^2 + \frac{\alpha_3^2}{\sin^2\theta} = \alpha_2^2, \qquad S_2 = -\int^\theta \left[\alpha_2^2 - \frac{\alpha_3^2}{\sin^2\theta}\right]^{\frac{1}{2}} d\theta; \tag{6.140}$$

$$\left.\begin{aligned} \alpha_1' &= \frac{1}{2m}\left[\left(\frac{\partial S_1}{\partial r}\right)^2 + \frac{\alpha_2^2}{r^2}\right] - \frac{Ze^2}{4\pi\varepsilon_0 r}, \\ S_1 &= \int^r \left[2m\alpha_1' + \frac{mZe^2}{2\pi\varepsilon_0 r} - \frac{\alpha_2^2}{r^2}\right]^{\frac{1}{2}} dr. \end{aligned}\right\} \tag{6.141}$$

The quantity α_1' is the energy, α_3 the angular momentum around the polar axis (the z-axis), while α_2 satisfies the equation

$$\alpha_2^2 = p_\theta^2 + \frac{\alpha_3^2}{\sin^2\theta} = p_\theta^2 + \frac{p_\varphi^2}{\sin^2\theta} = M^2, \tag{6.142}$$

where we have used (2.319). We see thus that α_2 is the total angular momentum. From (6.142) — and also from the physical meaning of α_2 and α_3 — it follows that $\alpha_2 \geqq \alpha_3$. The minus-sign in the second of equations (6.140) is introduced in order that β_2 and β_3 have a simple physical meaning [see (6.148) and (6.151)].

We shall now introduce instead of α_1' an α_1 which is related to α_1 by the equation

$$\alpha_1' = E = -Z^2e^4m/2(4\pi\varepsilon_0)^2\alpha_1^2. \tag{6.143}$$

The reason for this change-over will become apparent in the next section. We note that in assuming E to be negative we are restricting our discussion to elliptic orbits. We also note that α_1 has the dimensions of an angular momentum, that is, the same dimensions as α_2 and α_3,

Introducing a quantity R by the equation

$$R = Zme^2/4\pi\varepsilon_0, \tag{6.144}$$

we can write (6.141) in the form

$$S_1 = \int^r \left[\frac{\alpha_1^2-\alpha_2^2}{r^2} - \left(\frac{R}{\alpha_1} - \frac{\alpha_1}{r}\right)^2\right]^{\frac{1}{2}} dr. \tag{6.145}$$

As $\partial S_1/\partial r$ must always be real because of its physical meaning, it follows from (6.145) (i) that α_1 should be larger than or at least equal to α_2, and (ii) that if $\alpha_1 = \alpha_2$, the orbit is circular, since in that case we must have $R/\alpha_1 = \alpha_1/r$.

To find the physical meaning of the β_k we use equations (6.103). The Hamilton-Jacobi function is given by (6.137), (6.139), (6.140), and (6.145). We note, by the way, that now S_1, for instance, contains two α_k, as does S_2. We see from (6.143) that E contains only α_1 so that β_2 and β_3 are constants of motion. For β_3 we find

$$\beta_3 = \frac{\partial S}{\partial \alpha_3} = \int_{\pi/2}^{\theta} \frac{\alpha_3\, d\theta}{\sin\theta\sqrt{\alpha_2^2\sin^2\theta - \alpha_3^2}} + \varphi, \tag{6.146}$$

where we have chosen the lower limit of the integral occurring in S_2 to be equal to $\pi/2$. Let i be given by the equation

$$\alpha_3 = \alpha_2 \cos i, \tag{6.147}$$

from which it follows that i is the angle between the polar axis and the nor-

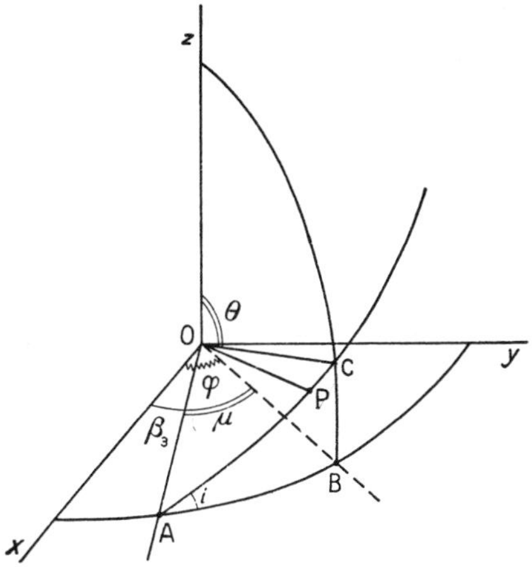

Fig. 26. The Kepler problem. OAC is the orbital plane; OA the line of the ascending node; OP the major axis; OC the radius vector of the point in the orbit; i the inclination of the orbital plane; and β_3 the length of the ascending node.

mal to the orbital plane (see fig. 26). We can then write (6.146) in the form

$$\begin{aligned}\beta_3 &= \int_{\pi/2}^{\theta} \frac{\cos i \, d\theta}{\sin\theta \sqrt{\sin^2\theta - \cos^2 i}} + \varphi \\ &= -\arc\sin(\cot i \cot\theta) + \varphi = -\mu + \varphi,\end{aligned} \tag{6.148}$$

where μ is the angle AOB in fig. 26. This follows from a consideration of fig. 27 which leads to

$$\cot i \cot \varphi = \frac{\mathrm{AB}}{\mathrm{BC}} \cdot \frac{\mathrm{BC}}{\mathrm{OB}} = \frac{\mathrm{AB}}{\mathrm{OB}} = \sin \mu. \tag{6.149}$$

We see thus that β_3 is the length of the ascending node (the angle xOA in fig. 26).

We now consider β_2 for which we have

$$\beta_2 = \frac{\partial S}{\partial \alpha_2} = \int^r \frac{-\alpha_2 dr/r^2}{\sqrt{\dfrac{\alpha_1^2 - \alpha_2^2}{r^2} - \left(\dfrac{R}{\alpha_1} - \dfrac{\alpha_1}{r}\right)^2}} - \int_{\pi/2}^{\theta} \frac{\alpha_2 \sin\theta \, d\theta}{\sqrt{\alpha_2^2 \sin^2\theta - \alpha_3^2}},$$

or, introducing the notation

$$a = \frac{\alpha_1^2}{R}, \qquad \frac{\alpha_2^2}{\alpha_1^2} = 1-\varepsilon^2, \qquad \frac{\alpha_2^2}{Rr} = 1+\varepsilon\cos\chi, \tag{6.150}$$

$$\beta_2 = \arcsin\,[\cos\theta/\sin i]-\chi = \psi-\chi, \tag{6.151}$$

where we have again used fig. 27; ψ is the angle AOC and χ the true anomaly (see fig. 28).

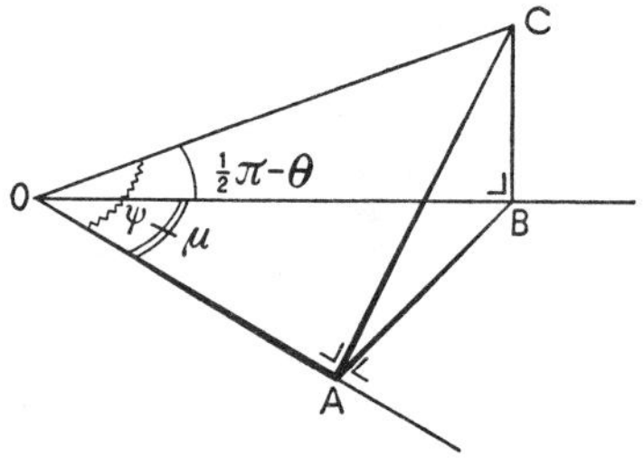

Fig. 27. The Kepler problem. The angles indicated by ∟ are right angles. The lettering is the same as in fig. 26.

The reason for introducing the last of equations (6.150) is that one can easily convince oneself that the radical occurring in S_1 is real only, if r lies between $a(1+\varepsilon)$ and $a(1-\varepsilon)$; this is, of course, related to our choice of a negative E, which corresponds to a finite orbit. The last of equations (6.150) shews, indeed, that this finite orbit is an elliptic one; we note also that ε must be less than or at most equal to unity as $\alpha_2 \leqq \alpha_1$. This last of equations (6.150) gives us the radius vector r as a function of the true anomaly.

As β_3 is a constant, the length of the ascending node is fixed. The fact that i is constant then fixes the orbital plane completely in space, and the fact that β_2 is constant together with the relation $\chi = \psi-\beta_2$, which follows from equation (6.151), then shews us that the orbit itself is fixed in space, that is, that the axes of the ellipse remain in a constant direction in space.

The equation for the orbit can be written in the form

$$\frac{a}{r} = \frac{1+\varepsilon\cos\chi}{1-\varepsilon^2}, \tag{6.152}$$

and we see that a is the semi-major axis and ε the eccentricity of the ellipse [compare the discussion of (1.237), (1.240), (1.243), and (1.244)].

We shall now consider β_1. To do this we shall first consider the orbit in somewhat more detail (see fig. 28). In terms of the Cartesian coordinates

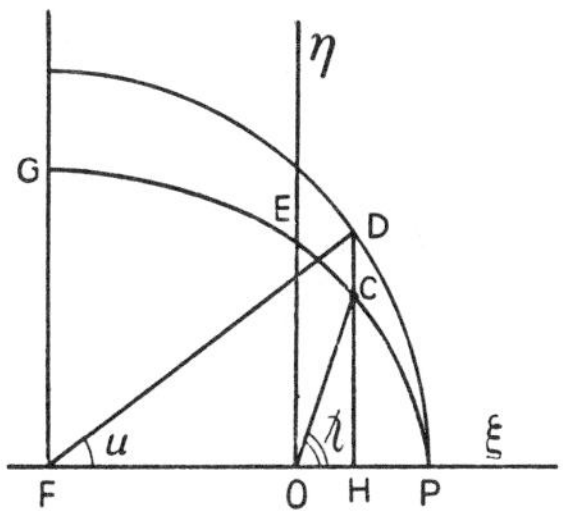

Fig. 28. The Kepler orbit. O is the origin which is the centre of the field of force; FP = a (the semi-major axis); OP = $a(1-\varepsilon)$ (the distance from the origin to the pericentre); OF = $a\varepsilon$; FG = $a\varepsilon = b$ (the semi-minor axis); OE = p (the parameter of the ellipse); χ is the true anomaly; and u the eccentric anomaly.

ξ and η we get the following parametric representation of the orbit

$$\begin{aligned} \xi &= r\cos\chi = a(\cos u-\varepsilon), \\ \eta &= r\sin\chi = a(1-\varepsilon^2)^{\frac{1}{2}}\sin u, \end{aligned} \tag{6.153}$$

where we have introduced the eccentric anomaly u by the equation

$$r = a(1-\varepsilon\cos u). \tag{6.154}$$

The angle u is the angle DFP in fig. 28, where CD is parallel to the η-axis. From fig. 28 and the fact that OF has the length $a\varepsilon$ and that the ratio of DH to CH is the same as that of the semi-major to the semi-minor axis, equations (6.153) follow easily, and as $\xi^2+\eta^2 = r^2$, equation (6.154) follows.

From (6.102) we have for β_1

$$\beta_1 = \gamma(t-\delta), \qquad \gamma = \frac{\partial \bar{H}}{\partial \alpha_1} = \frac{R^2}{m\alpha_1^3}, \tag{6.155}$$

while from (6.103) we get

$$\begin{aligned} \beta_1 = \frac{\partial S}{\partial \alpha_1} &= \int^r \frac{R^2}{\alpha_1^3} \frac{dr}{\left[\dfrac{\alpha_1^2-\alpha_2^2}{r^2} - \left(\dfrac{R}{\alpha_1} - \dfrac{\alpha_1}{r}\right)^2\right]^{\frac{1}{2}}} \\ &= \int_0^u (1-\varepsilon\cos u)du = u-\varepsilon\sin u. \end{aligned} \tag{6.156}$$

Combining (6.155) and (6.156) we get

$$u - \varepsilon \sin u = \gamma(t - \delta). \tag{6.157}$$

We see here the advantage of having introduced the eccentric anomaly: it is relatively simply related to the time, even though (6.157) is a transcendental equation. From (6.157) and (6.153) we can find the time-dependence of the true anomaly. Equations (6.153) which give the orbit in terms of Cartesian coordinates will be useful for the discussions in the next chapter.

We also want to write down an expression for the parameter p of the ellipse. This is the value of r when $\chi = \pi/2$, or, the intersection with the η-axis. From (6.152) and the third of equations (6.150) we have

$$p = a(1-\varepsilon^2) = \alpha_2^2/R; \tag{6.158}$$

we see that p is independent of α_1 and depends on α_2, that is, the total angular momentum only.

Let us briefly summarise the results of our calculations in so far as the physical meaning of the α_k and the β_k is concerned: α_1 determines the energy, or the semi-major axis [(6.143) and (6.150)]; α_2 is the total angular momentum [(6.142)] and together with α_1 determines the eccentricity [(6.150)]; α_3 is the component of the angular momentum along the polar axis [(6.139)] and together with α_2 determines the inclination of the orbital plane [(6.147)]; β_3 is the length of the ascending node [(6.148)]; β_2 determines the direction of the pericentre in the orbital plane [(6.151)]; and β_1 gives us, finally the relation between the eccentric anomaly and the time [(6.157)]. The quantity δ in (6.155) is the sixth and last constant of motion; its physical meaning is that it is the time of passage through the pericentre. The α_k and β_k are called the *elements* of the orbit.

2. ACTION AND ANGLE VARIABLES

We shall first of all consider a system with only one degree of freedom. Let its motion be periodic. This can occur in two different ways. Either, different values of q correspond to different states of the system, and both p and q are periodic functions of time: after a period τ both p and q return to the same value (*libration*; see fig. 29a); or, every time q increases by a certain constant amount q_0 the same state of the system is repeated: after a period τ p returns to the same value, but q is increased by q_0 (*rotation*; see fig. 29b). The same system can shew both libration and rotation. A simple pendulum, for instance, will for a small amplitude shew libration, but if the energy of the pendulum is sufficient to let it get round, it well shew rotation. Generally speaking, the first case usually occurs when the system moves between two states of vanishing kinetic energy, while in the second case one can always choose q to be an angle and we can choose its scale so that q_0 is equal to 2π.

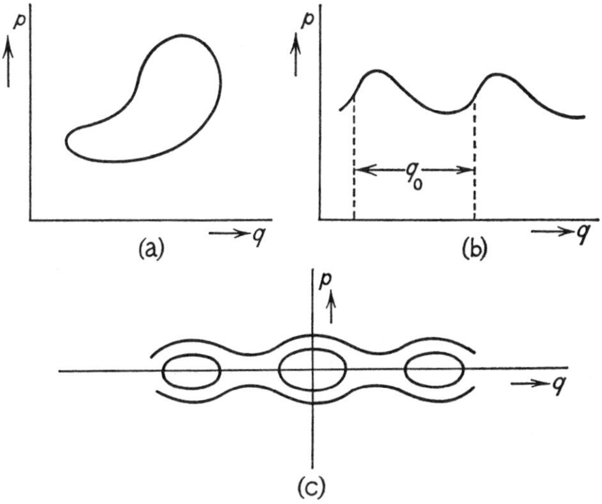

Fig. 29. The orbits in phase space corresponding to possible one-dimensional periodic motions: (a) libration, (b) rotation, (c) orbits for a simple pendulum shewing both libration and rotation.

Let us assume that we have found a solution of the Hamilton-Jacobi equation for our system, and that α and β are the resultant new canonical coordinates. We know that then β will be a linear function of the time t [see (6.102)],

$$\beta = \gamma(t - t_0). \tag{6.201}$$

We then have the following two cases:

$$\text{libration:} \qquad q(\beta+\gamma\tau) = q(\beta), \tag{6.202a}$$

$$\text{rotation:} \qquad q(\beta+\gamma\tau) = q(\beta)+2\pi. \tag{6.202b}$$

We can now make a transformation from α and β to a new set of canonical variables J and w such that

$$w = \beta/\gamma\tau. \tag{6.203}$$

Equation (6.203) is a 'point' transformation of the type (5.221 iv) or (5.203) and corresponds thus to a canonical transformation. Indeed, it is generated by the function

$$S' = J\beta/\gamma\tau, \tag{6.204}$$

from which follow (6.203) and the equation

$$J = \alpha\gamma\tau. \tag{6.205}$$

In terms of w we get instead of equations (6.202)

$$\text{libration:} \qquad q(w+1) = q(w), \tag{6.206a}$$

$$\text{rotation:} \qquad q(w+1) = q(w)+2\pi. \tag{6.206b}$$

The canonical transformation from p and q to J and w will be a Hamilton-Jacobi transformation, as the transformed Hamiltonian was a function of α only and as J does not contain β, so that $\bar{H}$ will be a function of J only. Let $\bar{S}(q, J)$ be the Hamilton-Jacobi function which generates this transformation, so that

$$p = \frac{\partial \bar{S}}{\partial q}, \qquad w = \frac{\partial \bar{S}}{\partial J}. \tag{6.207}$$

From these equations it follows that

$$\frac{\partial w}{\partial q} = \frac{\partial}{\partial J}\frac{\partial \bar{S}}{\partial q} = \frac{\partial p}{\partial J}, \tag{6.208}$$

or,

$$\frac{\partial}{\partial J}\oint p\,\mathrm{d}q = \frac{\partial}{\partial J}\oint \frac{\partial \bar{S}}{\partial q}\,\mathrm{d}q = \oint \frac{\partial w}{\partial q}\,\mathrm{d}q = \oint \mathrm{d}w = 1, \tag{6.209}$$

where $\oint$ indicates integration over one complete period, and where we have used the fact that w was chosen in such a way that it increases by unity over one period [see (6.206)]. From (6.209) it follows that

$$J = \oint p\,\mathrm{d}q, \tag{6.210}$$

and from the first of equations (6.207) and (6.210) we see that J is the increase of $\bar{S}$ over one period. We also note that J is just the action integral taken over one period. As J has the dimensions of an angular moment, w has the dimensions of an angle; hence J and w are known as *action* and *angle variables*.

The Hamiltonian is now a function, $E(J)$, of J only and we get from the canonical equations of motion

$$J = \text{constant}, \quad \text{and} \quad \dot{w} = \frac{\partial E}{\partial J} = \nu, \tag{6.211}$$

or

$$w = \nu(t-t_0). \tag{6.212}$$

Using the fact that w increases by unity over a period, we get

$$\Delta w = 1 = \nu\tau, \quad \text{or} \quad \nu = 1/\tau, \tag{6.213}$$

so that $\nu = \partial E/\partial J$ is the frequency of the motion. This shows that we can obtain the frequency (or period) of the motion as soon as we know the Hamiltonian as a function of the action variable J without having to solve the equations of motion [compare (5.409)].

We could have introduced J by (6.210) without first discussing equations (6.202) and we could thus have proved that, indeed, w changes by unity over one period. We shall use that approach for a discussion of system of several degrees of freedom. The only systems which we shall discuss are those which are multiply periodic, that is, which are periodic in each of their q_i, and for which the Hamilton-Jacobi equation can be solved by a separation of variables so that

$$S(q, \alpha) = \sum_i S_i(q_i; \alpha_k). \tag{6.214}$$

The new (action) variables J_i are introduced by the equations

$$J_i = \oint p_i \mathrm{d}q_i, \tag{6.215}$$

where each integral is over the period corresponding to the q_i it involves; these periods are not necessarily all the same. We notice that all J_i have the dimensions of an angular momentum or action. We also note that, if q_i is cyclic so that p_i is a constant [see (2.402)], it follows from (6.215) that then $J_i = 2\pi p_i$.

As the integration in (6.215) is over q_i, and as q_i is the only one of the q_k which occurs in the integral on the right hand side of equation (6.215) since the Hamilton-Jacobi equation was separable, the J_i will be functions of the α_k only and will not contain the β_k. This means that the transformation from the p_k and q_k to the J_i and w_i will be a Hamilton-Jacobi transformation, leading to a transformed Hamiltonian $\overline{H}$ which is a function of the J_i only.

Let this Hamilton-Jacobi transformation be generated by the Hamilton-Jacobi function $\bar{S}$. From the fact that the original Hamilton-Jacobi equation was separable and the fact that the J_i are functions of the α_k only, it follows that we can write for $\bar{S}$

$$\bar{S}(q, J) = \sum_i \bar{S}_i(q_i; J_k), \tag{6.216}$$

and using the first of the transformation equations

$$p_i = \frac{\partial \bar{S}}{\partial q_i} = \frac{\partial \bar{S}_i}{\partial q_i}, \qquad w_i = \frac{\partial \bar{S}}{\partial J_i}, \tag{6.217}$$

we see from (6.215) that the J_i are equal to the increase of S_i over one period.

From the canonical equations of motion we get

$$\dot{w}_i = \frac{\partial \bar{H}}{\partial J_i} = \nu_i(J_1, \ldots, J_s), \quad \text{or,} \quad w_i = \nu_i t + \delta_i. \tag{6.218}$$

One can prove as follows that the ν_i are, indeed, the frequencies of the motion. Let q_j go through its period while the other q's are kept fixed. We then get for the increase in w_i (the index j on Δ and δ indicates that only q_j is changed)

$$\begin{aligned} \Delta_j w_i &= \oint \delta_j w_i = \oint \frac{\partial w_i}{\partial q_j} \mathrm{d}q_j = \oint \frac{\partial^2 \bar{S}}{\partial q_j \partial J_i} \mathrm{d}q_j \\ &= \frac{\partial}{\partial J_i} \oint p_j \mathrm{d}q_j = \frac{\partial J_j}{\partial J_i} = \delta_{ij}. \end{aligned} \tag{6.219}$$

Hence, if τ_i is the period corresponding to q_i we get from (6.218) and (6.219)

$$\Delta_i w_i = \nu_i \tau_i = 1, \tag{6.220}$$

which proves our statement.

To conclude this section we shall consider two examples: the one-dimensional harmonic oscillator and the Kepler problem. From (6.118), (6.121), and (6.210) we get

$$\begin{aligned} J &= \oint [2m\alpha - maq^2]^{\frac{1}{2}} \mathrm{d}q \\ &= 2\alpha(m/a)^{\frac{1}{2}} \int_0^{2\pi} \cos^2\theta \, \mathrm{d}\theta = 2\pi\alpha(m/a)^{\frac{1}{2}}, \end{aligned} \tag{6.221}$$

where we have used the substitution $q = (2\alpha/a)^{\frac{1}{2}} \sin\theta$.

From (6.221) and (6.115) it follows that

$$E = \alpha = (J/2\pi)(a/m)^{\frac{1}{2}}, \tag{6.222}$$

and from (6.211) that

$$\nu = \partial E/\partial J = (2\pi)^{-1}(a/m)^{\frac{1}{2}}. \tag{6.223}$$

We see that we have, indeed, obtained the frequency of the motion without solving the equations of motion.

In the Kepler problem, we can shew that J_1, J_2, and J_3 are linear combinations of α_1, α_2, and α_3. From (6.215) we get, using (6.139), (6.142), and (6.141)

$$J_3 = \oint p_\varphi \mathrm{d}\varphi = \oint \alpha_3 \mathrm{d}\varphi = 2\pi\alpha_3, \tag{6.224}$$

$$J_2 = \oint p_\theta \mathrm{d}\theta = \oint \left[\alpha_2^2 - \left(\frac{\alpha_3}{\sin\theta}\right)^2\right]^{\frac{1}{2}} \mathrm{d}\theta = 2\pi(\alpha_2 - \alpha_3), \tag{6.225}$$

$$J_1 = \oint p_r \mathrm{d}r = \oint \left[2mE + \frac{Zme^2}{2\pi\varepsilon_0 r} - \frac{\alpha_2^2}{r^2}\right]^{\frac{1}{2}} \mathrm{d}r$$

$$= -2\pi\alpha_2 + \frac{Ze^2}{4\varepsilon_0}\sqrt{\frac{2m}{-E}} = 2\pi(\alpha_1 - \alpha_2). \tag{6.226}$$

From (6.226) it follows that the energy E expressed in terms of J_1, J_2, and J_3 is given by the equation

$$E = \frac{-Z^2me^4}{8\varepsilon_0^2(J_1+J_2+J_3)^2}. \tag{6.227}$$

We see that in this case $\nu_1 = \nu_2 = \nu_3$. This is not surprising as we are dealing with a closed orbit.

From (6.226) we see also one of the advantages of using α_1 instead of α_1'. The advantages of using the J_k will become apparent in the next section; we only draw attention here to the fact that they are the quantities which in the old quantum theory were quantised through the Sommerfeld-Wilson quantisation rules.

Equations (6.225) and (6.226) can be derived by straightforward integration, but there are more elegant ways of obtaining the same result. The easiest way to evaluate J_2 is to notice that $p_r\dot{r}+p_\theta\dot{\theta}+p_\varphi\dot{\varphi}$ is twice the total kinetic energy of the particle split up into three contributions corresponding to motion in the r-, the θ-, and the φ-direction, respectively. If one splits the kinetic, energy however, up into three contributions corresponding to the radial motion, the transverse motion in the orbital plane, and the motion perpendicular to the orbital plane (which does not contribute to the kinetic energy), the result is

$$p_r\dot{r}+p_\theta\dot{\theta}+p_\varphi\dot{\varphi} = p_r\dot{r}+p_\chi\dot{\chi}+0, \tag{6.228}$$

where χ is again the true anomaly; as p_χ is the total angular momentum M we have

$$p_\theta\dot{\theta} = M\dot{\chi}-p_\varphi\dot{\varphi} = \alpha_2\dot{\chi}-\alpha_3\dot{\varphi}, \tag{6.229}$$

and (6.225) follows when we use the fact that the orbit is a closed one, which means that when θ goes through its period — which is from $\pi/2-i$ to $\pi/2+i$ and back — χ and φ increase by 2π.

To evaluate J_1 one can introduce a new variable u by the equation

$$2r = (r_1+r_2)+(r_1-r_2)u, \tag{6.230}$$

where r_1 and r_2 are the maximum and minimum values of r. A more elegant method which is due to Born — and which can also be used to evaluate J_2 — consists in considering the integral over r to be an integral in the complex plane (see fig. 30). The integrand p_r is a two-valued function and we must

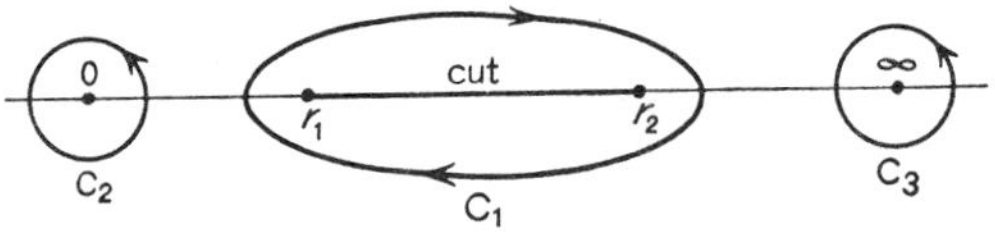

Fig. 30. The contours in the complex r-plane involved in the evaluation of J_1.

therefore introduce a cut in the r-plane between the two branch points r_1 and r_2. In fig. 30 we have chosen the branches such that the positive square root is taken above the real axis and the negative square root beneath it. We then have

$$J_1 = \oint_{C_1} p_r \mathrm{d}r. \tag{6.231}$$

The contour C_1 encloses the cut, but by distorting it into C_2+C_3 we are left to deal with contour integrals which can be evaluated by Cauchy's theorem of residues. The two poles of the integrand are at $r=0$ and at $r=\infty$, so that we have

$$J_1 = \oint_{C_2} p_r \mathrm{d}r + \oint_{C_3} p_r \mathrm{d}r$$
$$= 2\pi\mathrm{i}\,[\text{Residue at } r = 0 + \text{Residue at } r = \infty]. \tag{6.232}$$

The first residue is easily seen to be equal to $\mathrm{i}\alpha_2$, and the second one can be evaluated by introducing r^{-1} as a new variable and expanding in a power series in r^{-1} whence follows the result of (6.226).

3. ADIABATIC INVARIANTS

In the preceding section we saw that we could choose sets of canonically conjugate variables in such a way (*i*) that the Hamiltonian is a function of only half of the variables and (*ii*) that for periodic systems for which the Hamilton-Jacobi equation can be solved by separating the variables one can choose angle variables such that they change by unity over one period. The reasons for introducing variables such that the Hamiltonian depends on one half of the variables only are obvious, but the reasons for introducing the action and angle variables are much more sophisticated. Indeed, they only came to the fore when the old quantum theory was developed, and the reason was that the action variables are so-called *adiabatic invariants*. We shall define an adiabatic invariant as a quantity which will remain invariant when parameters occurring in the Hamiltonian are changed slowly. We shall make this definition more quantitative in a moment, and we shall shew that, indeed, the action variables are adiabatic invariants.

We wish to say a few words about the importance of adiabatic invariants and the reasons why they were introduced in the old quantum mechanics. It was pointed out by Ehrenfest that if one wanted to find suitable quantities to quantise, they should be adiabatic invariants. The reason for this is that if a parameter is changed so slowly that in its Fourier expansion there are only frequencies which are below a certain limit, ν_0 say, which is less than any frequency corresponding through the Bohr conditions to a quantum transition, no quantum transition can take place during the variation of the parameter. This, in turn, means that during such a slow variation of parameters occurring in the Hamiltonian the quantum numbers cannot change, and the quantities which are quantised can therefore not change either. As the action variables are adiabatic invariants, they are suitable objects for quantisation, and they are, indeed, quantised by the Sommerfeld-Wilson quantisation rules. The reason for calling such slow changes adiabatic are that one shews in statistical mechanics that the entropy of a system is determined by the distribution of the constituent parts of the system over possible energy states. As there cannot be any transitions to other states during an adiabatic change of parameters, the entropy will not change: this corresponds to the thermodynamic definition of an adiabatic change. We may mention here that adiabatic invariance also plays an important role in modern quantum theory [see, for instance, H. A. Kramers, Quantum Theory (North Holland Publishing Company, Amsterdam, 1957), p. 216] where it states that a system which is in a stationary state will stay in this stationary state during adiabatic processes.

Recently there has been renewed interest in adiabatic invariants, as they play an important role both in the theory of particle accelerators and in the theory of the motion of charged particles in magnetic fields such as those proposed for thermonuclear experiments.

For the sake of simplicity we shall prove the adiabatic invariance of the action variables only for the one-dimensional case; we may remark that a similar proof holds also for the case of more degrees of freedom, as long as there is no relation of the kind

$$\sum_i \nu_i k_i = 0, \tag{6.301}$$

where the k_i are positive or negative integers, between the frequencies ν_i defined by (6.218). A system for which at least one relation of the kind (6.301) exists is said to be degenerate, and such systems will be excluded from the discussion in the present section. Such systems often occur in nature, and the Kepler problem deals with such a system: we noticed in the preceding section that it followed from (6.227) that $\nu_1 = \nu_2 = \nu_3$.

We are concerned now with a system described by a Hamiltonian

$$H = H(p, q; a(t)), \tag{6.302}$$

where $a(t)$ is a parameter which changes with time. In the limit where

$$\dot{a}(t) \to 0 \tag{6.303}$$

we say that we are dealing with an adiabatic change. To perform the transformation from p, q to the action and angle variables J, w, we shall use a generating function of the kind (5.220a), $W(q, w, a)$, which will now depend on the time through the parameter a. The transformation is given by the equations

$$p = \frac{\partial W}{\partial q}, \qquad J = -\frac{\partial W}{\partial w}. \tag{6.304}$$

As the Hamiltonian in this case depends on the time, we must proceed with some care. We shall use the fact, proved in § 5.4, that the canonical equations of motion are equivalent to the variational principle (5.425). This means that from the equation

$$\delta \int_1^2 [p\dot{q} - H(p, q, a)]\mathrm{d}t = 0 \tag{6.305}$$

the equation

$$\delta \int_1^2 [J\dot{w} - \bar{H}(J, w, a)]\mathrm{d}t = 0 \tag{6.306}$$

must follow since the transformation (6.304) is a canonical one. It follows that

$$p\dot{q} - H(p, q, a) = J\dot{w} - \bar{H}(J, w, a) + \frac{\mathrm{d}}{\mathrm{d}t} V(q, w, a), \tag{6.307}$$

and using equations (6.304) and comparing the coefficients of $\dot{q}$ and $\dot{w}$ on both sides of this equation, we see that

$$V \equiv W, \tag{6.308}$$

and also that

$$\bar{H}(J, w, a) = H(J, a) + \frac{\partial W}{\partial t}, \tag{6.309}$$

where we have taken into account that after the transformation H will not contain w. The equations of motion for w and J are now

$$\dot{w} = \frac{\partial \bar{H}}{\partial J} = \frac{\partial H}{\partial J} + \frac{\partial}{\partial J}\frac{\partial W}{\partial t} = \frac{\partial H}{\partial J} + \frac{\partial}{\partial J}\left(\frac{\partial W}{\partial a}\dot{a}\right), \tag{6.310}$$

$$\dot{J} = -\frac{\partial \bar{H}}{\partial w} = -\frac{\partial H(J, a)}{\partial w} - \frac{\partial}{\partial w}\frac{\partial W}{\partial t} = -\frac{\partial}{\partial w}\left(\frac{\partial W}{\partial a}\dot{a}\right). \tag{6.311}$$

Let us now assume that J has a given value $J(0)$ at $t = 0$, and investigate the change in its value during a time interval T, during which a changes from its original value a_0 to a value $a_0 + \delta a$. To simplify our discussion we shall assume that the change in a takes place linearly, which means that $\dot{a}$ (which is now equal to $\delta a/T$) is a constant, and also that $\dot{a}$ will tend to zero at the same time as we let T tend to infinity, while $\dot{a}T$ will remain finite. From (6.311) we get for the change in J

$$J(T)-J(0) = -\int_0^T \frac{\partial}{\partial w}\frac{\partial W}{\partial a}\,\dot{a}\,\mathrm{d}t \tag{6.312a}$$

$$= -\dot{a}\int_0^T \frac{\partial}{\partial w}\frac{\partial W}{\partial a}\,\mathrm{d}t \tag{6.312b}$$

$$= -\dot{a}\int_0^T \sum_{k\neq 0} A_k(J, a)\mathrm{e}^{2\pi i k w}\mathrm{d}t \tag{6.312c}$$

$$= -\dot{a}\int_0^T \sum_{k\neq 0} (A_k^{(0)}+A_k^{(1)}\dot{a}t+\ldots)$$
$$\times \exp\,[2\pi\mathrm{i}(\nu_0 t+\delta_0+(\nu_1 t^2+\delta_1 t)\dot{a}+\ldots)]\mathrm{d}t \tag{6.312d}$$

$$= -\dot{a}\int_0^T \left\{\sum_{k\neq 0} A_k^{(0)}\exp\,[2\pi i k(\nu_0 t+\delta_0)]\mathrm{d}t\right\}$$
$$+\dot{a}^2\int_0^T (\ldots)\mathrm{d}t+\ldots. \tag{6.312e}$$

In going over from (6.312a) to (6.312b) we have used the fact that $\dot{a}$ is assumed to be constant. Equation (6.312c) follows from the fact that our system is periodic in w with a period of unity so that we can expand W in a Fourier series. The variable w is, however, no longer a strictly linear function of the time, but we have from (6.310)

$$w = \frac{\partial H}{\partial J}\,t+\delta(J, a) = \nu(J, a)t+\delta(J, a)$$

$$= \nu_0 t+\delta_0+(\nu_1 t^2+\delta_1 t)\dot{a}+\ldots, \tag{6.313}$$

where $\delta(J, a)$ would be strictly constant, were a constant, and where we have expanded ν and δ in power series in t.

We can now let T go to infinity. As the modulus of the integral in the first term has as its upper limit the value $\sum_{k\neq 0}\,[A_k^{(0)}/2\pi k\nu_0]$ which is finite, the first term tends to zero as $\dot{a}$ tends to zero. The second term is at most of the order $\dot{a}^2 T = (\dot{a}T)\dot{a}$ which also tends to zero as $\dot{a}$ tends to zero. We see thus that in the limit as $\dot{a}$ tends to zero, J remains unchanged. We do not wish to discuss in any detail the more general case where the higher derivatives of a

do not vanish, but refer the reader to the literature [for instance, J. M. Burgers, Ann. Phys. (Lpz) **52** (1917) 195].

We wish to draw briefly attention to the connexion between the invariance of J and Liouville's theorem in statistical mechanics which states the invariance of volume elements in phase space. To see this connexion for the one-dimensional case which we have been considering we write

$$J = \oint p \mathrm{d}q = \iint \mathrm{d}p \mathrm{d}q = \iint \mathrm{d}\Omega,$$

where $\mathrm{d}\Omega$ is an element in phase space.

To conclude this section we shall consider two examples. The first one is the case of the simple pendulum which we shall consider in the small amplitude approximation so that its equation of motion is the same as that of a linear harmonic oscillator. The second example is the case of a charged particle in a magnetic field.

Let l be the length of the pendulum, θ its angular displacement, q its linear displacement, ν its frequency, E its energy, m the mass of the bob, and

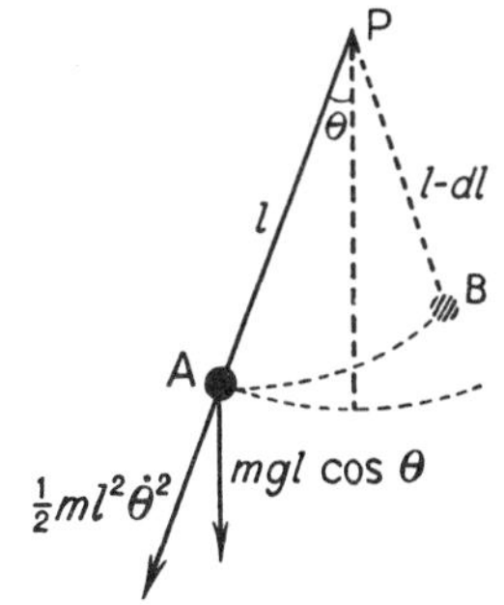

Fig. 31. The change in motion of a simple pendulum when its length is changed adiabatically from l to $l-\mathrm{d}l$. P is the point of suspension.

g the gravitational acceleration. The question to be put now is: How will θ change when l is changed adiabatically. The answer to this question can be given in two ways. The first method is to consider in a straightforward fashion the changes in the mechanical system when l is adiabatically changed from l to $l+\mathrm{d}l$. The second — and much quicker — method is to use the adiabatic invariance of J. We shall consider both methods.

The work $\mathrm{d}W$ done when l is changed consists of two parts: the work done against gravity and the work done against the centripetal forces, or,

$$\mathrm{d}W = -mg\,\overline{\cos\theta}\,\mathrm{d}l - ml\overline{\dot{\theta}^2}\,\mathrm{d}l, \tag{6.314}$$

where a bar denotes an average taken over the period during which l changes. After the pendulum's length has attained its new length we can strike an energy balance and find

$$\mathrm{d}W = -mg\,\mathrm{d}l + \mathrm{d}E, \tag{6.315}$$

where $\mathrm{d}E$ is the change in the energy of the pendulum and $-mg\,\mathrm{d}l$ the potential energy change because of the altered position of the swing of the pendulum bob in the gravitational field. Combining (6.314) and (6.315) we get, putting $\cos\theta \doteqdot 1-\frac{1}{2}\theta^2$,

$$\mathrm{d}E = \tfrac{1}{2}mg\overline{\theta^2}\,\mathrm{d}l - ml\overline{\dot{\theta}^2}\,\mathrm{d}l. \tag{6.316}$$

As the change in length is a slow one, that is, takes place over many oscillations of the pendulum, we have

$$\tfrac{1}{2}E = \tfrac{1}{2}ml^2\overline{\dot{\theta}^2} = \tfrac{1}{2}mgl\overline{\theta^2}, \tag{6.317}$$

and (6.316) can be rewritten in the form

$$\mathrm{d}E = -(E/2l)\,\mathrm{d}l. \tag{6.318}$$

It follows that E^2l is invariant, and since E is proportional to $l\theta_0^2$ [compare (6.317)] where θ_0 is the amplitude of the oscillation, we see that

$$l^3\theta_0^4 = \text{invariant}. \tag{6.319}$$

If we use the adiabatic invariance of J, the proof of (6.319) is simple. From (6.222) and (6.223) we have

$$J = E/\nu, \tag{6.320}$$

and since E is proportional to $l\theta_0^2$ while for a simple pendulum ν is proportional to $l^{-\frac{1}{2}}$, (6.319) follows at once. We may draw attention to the fact that combining the fact that ν is proportional to $l^{-\frac{1}{2}}$ with (6.318) leads immediately to a proof for the invariance of E/ν, that is, of J for this particular case.

In conclusion we wish to say a few words about the case of a charged particle in a magnetic field. We shall consider only the most important of the adiabatic invariants, and we shall restrict ourselves to the simple case where the magnetic field is a uniform field in the z-direction which is produced by a vector potential A with x, y, z-components $-\frac{1}{2}By$, $\frac{1}{2}Bx$, 0. The action variable to be considered is

$$J_\theta = \oint p_\theta \mathrm{d}\theta, \tag{6.321}$$

where we have introduced cylindrical coordinates r, θ, z for the moment. Expressed in cylindrical coordinates the Lagrangian of our system is [compare (5.353)]

$$L = \tfrac{1}{2}m(\dot{r}^2 + r^2\dot{\theta}^2 + \dot{z}^2) + \tfrac{1}{2}eBr^2\dot{\theta}, \tag{6.322}$$

and we have thus for p_θ the equation

$$p_\theta = mr^2\dot{\theta} + \tfrac{1}{2}eBr^2. \tag{6.323}$$

To simplify our considerations even further, we shall consider the case where $\dot{z} = 0$ and $\dot{r} = 0$, that is, the case where the particle is circling around the lines of force with the cyclotron frequency ω_c,

$$\omega_c = eB/m, \tag{6.324}$$

so that $\dot{\theta} = \omega_c$. From (6.321), (6.323), and (6.324) we then get

$$J_\theta = (6\pi m/e)(mv_\perp^2/2B) = (6\pi m/e)\mu \tag{6.325}$$

where $v_\perp$ is the transverse velocity

$$v_\perp = r\dot{\theta}, \tag{6.326}$$

and μ the magnetic moment corresponding to the motion of the particle,

$$\mu = \tfrac{1}{2}ev_\perp r = ev_\perp^2/2\omega_c = mv_\perp^2/2B. \tag{6.327}$$

We have thus for our special simplified case proved that the magnetic moment μ is an adiabatic invariant. This result remains valid for more complicated motions of a charged particle and also in inhomogeneous magnetic fields. Apart from μ there are two other adiabatic invariants. If one considers the motion of charged particles in a magnetic field configuration such as the earth's magnetic field or the field in a thermonuclear, magnetic mirror machine, one can prove [see Northrop and Teller, Phys. Rev. **117** (1960) 215] that the particles will remain trapped in the magnetic field, unless the adiabatic conditions are violated (assuming the absence of scattering which introduces another source of loss).

Chapter 7

PERTURBATION THEORY

In this chapter some methods are described for dealing with systems, the equations of motion of which cannot be solved rigorously, but which are such that a simplified problem — called the unperturbed problem — can be solved, while the difference between the original, perturbed system and the simplified, unperturbed system can be considered to be a small perturbation. In the first section we consider straightforward methods to treat perturbations and we apply these methods to the anharmonic oscillator problem. In the second section we discuss canonical perturbation theory — on which quantum mechanical perturbation theory is based. The question of secular and periodic perturbations is also discussed briefly. In the last section, finally, we consider the influence of weak electric and magnetic fields on the motion of charged particles and, in particular, on the motion of the hydrogen atom.

1. THE ANHARMONIC OSCILLATOR

The case where one is able to find an exact solution of the equations of motion of an actual physical system is an exception rather than the rule. The reasons for this are many. In the preceding chapters we have usually considered problems which could be reduced to relatively simple single-particle equations. Many of these single-particle problems involved central forces which were shewn to be reducible to a quadrature [see (1.219)]. The problems which we considered were also usually such that the quadratures led to solutions in closed form. These simple single-particle equations are for most systems only a first approximation to the actual equations of motion, and are obtained by neglecting 'perturbing' influences. These perturbations may be of several different kinds. First of all, there is the case where an unperturbed system is put in an external field, such as an external electric or magnetic field. This leads to the Stark effect and the Zeeman effect which we shall discuss in the last section of this chapter. Secondly, there is the case where in the discussion of the unperturbed system we have neglected the influence of parts of the system. As an example, we may mention the motion of the moon around the earth. To a first approximation, we consider both the moon and the earth to be point particles moving in orbits determined solely by the gravitational forces between the two masses. This solution must, however, be corrected both for the influence of the sun on the moon's orbit, and also for the fact that the earth is not a rigid body, but is highly deformable because it is covered by the oceans in which tides occur. We shall not discuss this kind of problem here — it is more suitably discussed in a textbook on celestial mechanics. Thirdly, there are the cases where we are presented with a system the equations of motion of which are too complicated to allow an exact solution in closed form; often, however, it is possible to find a system with a Hamiltonian which is very nearly the same as the Hamiltonian of the original system, but which allows us to solve the equations of motion by quadrature in closed form. The difference between the original and the simplified Hamiltonians can then be considered to be the 'perturbation' in this case. The anharmonic oscillator belongs to this class. This is a case which occurs in the theory of small vibrations, discussed in Chapter 3. In that chapter we retained only the first non-trivial terms in the potential energy which led to equations of motion which could be reduced to those of a set of independent harmonic oscillators. This is the unperturbed system. The perturbation consists of all the other terms in the Hamiltonian. The most important ones are the cubic terms. We shall not discuss here the general problem of a system of many degrees of freedom which to a first approximation reduces to the small vibrations problem of Chapter 3, but we shall discuss in some detail the one-dimensional anharmonic oscillator, the Hamiltonian of which is given by the equation

$$H = \frac{p^2}{2m} + \tfrac{1}{2}m\omega^2 q^2 + \lambda q^3. \tag{7.101}$$

The reason why it is feasible to consider the simplified unperturbed problems is that these are sufficiently similar to the original problem so that their solutions bear at least some relation to the required solutions of the actual problems. Moreover, one can usually find a solution of the perturbed problem in terms of a power series in some parameter which is contained as a factor in the perturbation — such as λ in equation (7.101) — and one hopes that, as long as λ is sufficiently small, the first few terms of this power series represent a good approximation to the solution of the perturbed problem.

In the next section we shall develop a systematic theory for these problems, based upon the use of the action and angle variables introduced in the preceding chapter. It may be asked whether it is really necessary — apart from the close connexion with quantum mechanical perturbation theory — to use all the paraphernalia of canonical theory, and in fact some authors have suggested that any straightforward method would do just as well. One could counter this argument by pointing to the fact that canonical perturbation theory was used extensively long before the advent of quantum mechanics, but a more convincing argument is probably to shew that in many cases straightforward methods are either more cumbersome, or lead to the wrong result, or both. To do this we shall consider in the present section two straightforward methods. These methods will be applied to the anharmonic oscillator problem with the Hamiltonian (7.101). The results will be compared with the solution obtained by canonical perturbation theory in the next section.

The problem to be considered in this chapter is that of a system the Hamiltonian H of which is of the form

$$H = H_0 + \lambda H_1, \tag{7.102}$$

where H_0 is the Hamiltonian of the unperturbed system, and λH_1 the perturbation. The canonical equations of motion which must be solved are [see (5.108)]

$$\dot{p}_k = -\frac{\partial H}{\partial q_k}, \qquad \dot{q}_k = \frac{\partial H}{\partial p_k}. \tag{7.103}$$

The first method of solving these equations is to put

$$\begin{aligned} p_k &= p_k^{(0)} + \lambda p_k^{(1)} + \lambda^2 p_k^{(2)} + \dots, \\ q_k &= q_k^{(0)} + \lambda q_k^{(1)} + \lambda^2 q_k^{(2)} + \dots. \end{aligned} \tag{7.104}$$

The $p_k^{(0)}$ and $q_k^{(0)}$ are the solutions of the unperturbed problem

$$\dot{p}_k^{(0)} = -\left(\frac{\partial H_0}{\partial q_k}\right)_0, \qquad \dot{q}_k^{(0)} = \left(\frac{\partial H_0}{\partial p_k}\right)_0, \tag{7.105}$$

where the subscript '0' indicates that we must substitute for p_k and q_k their unperturbed values $p_k^{(0)}$ and $q_k^{(0)}$. The $p_k^{(1)}$ and $q_k^{(1)}$ are found by substituting (7.102) and (7.104) into (7.103), and using (7.105). This gives us the following equations

$$\dot{p}_k^{(1)} = -\left(\frac{\partial H_1}{\partial q_k}\right)_0 - \sum_l \left(\frac{\partial^2 H_0}{\partial q_k \partial q_l}\right)_0 q_l^{(1)} - \sum_l \left(\frac{\partial^2 H_0}{\partial q_k \partial p_l}\right)_0 p_l^{(1)}, \tag{7.106}$$

$$\dot{q}_k^{(1)} = \left(\frac{\partial H_1}{\partial p_k}\right)_0 + \sum_l \left(\frac{\partial^2 H_0}{\partial p_k \partial q_l}\right)_0 q_l^{(1)} + \sum_l \left(\frac{\partial^2 H_0}{\partial p_k \partial p_l}\right)_0 p_l^{(1)}. \tag{7.107}$$

One can in this way find as many terms of the series (7.104) as one needs.

The second method is applicable only to multiply periodic systems where one can use the method of separation of variables (see § 6.2). In that case one can use a Hamilton-Jacobi function of the kind

$$S = \sum_k S_k(q_k;\, \alpha_i;\, \lambda), \tag{7.108}$$

where each S_k depends on one coordinate, q_k, only, but may contain all the α_i ($i = 1, \ldots, s$; s : number of degrees of freedom). We have indicated in equation (7.108) that the S_k will be functions of the perturbation parameter λ. We can now use the S_k to introduce the action variables J_k by the equations

$$J_k = \oint \frac{\partial S_k}{\partial q_k}\, dq_k, \tag{7.109}$$

where the integration is over one complete period of the q_k. The energy E of the system is a function of the α_i, and as the J_k are now through equations (7.109) functions of the α_i, we can find E as a function of the J_k. In practical cases, one expands the S_k and the J_k in power series in λ. One must take into account the fact that the periods of the q_k are also different in the perturbed system.

If one wants to obtain the change in the q_k one must evaluate the angle variables as well using the equations [compare (6.208)]

$$w_k = \oint \frac{\partial^2 S_k}{\partial q_k \partial J_k}\, dq_k, \tag{7.110}$$

and the equations which govern the transformation from the p_k and q_k to the w_k and J_k.

Of course, if all equations can be solved exactly in closed form, there is no need for perturbation theory, but one is usually dealing with systems where this cannot be done and where one therefore must use power series expansions in λ to find solutions which can be used.

We shall now apply the methods discussed here to the system governed by the Hamiltonian (7.101). We have thus

$$H_0 = \frac{p^2}{2m} + \tfrac{1}{2}m\omega^2 q^2, \qquad H_1 = q^3. \tag{7.111}$$

Using the first method one gets a zeroth order equation

$$m\ddot{q}^{(0)} + m\omega^2 q^{(0)} = 0, \tag{7.112}$$

with the solution

$$q^{(0)} = A \sin \omega t; \tag{7.113}$$

a first order equation

$$m\ddot{q}^{(1)} + m\omega^2 q^{(1)} = -3(q^{(0)})^2 = -3A^2 \sin^2 \omega t, \tag{7.114}$$

with the solution

$$q^{(1)} = -\frac{A^2}{2m\omega^2}(3 + \cos 2\omega t); \tag{7.115}$$

and a second order equation

$$m\ddot{q}^{(2)} + m\omega^2 q^{(2)} = -6q^{(0)}q^{(1)} = \frac{3A^3}{m\omega^2} \sin \omega t(3 + \cos 2\omega t), \tag{7.116}$$

with the solution

$$q^{(2)} = \frac{A^3}{m^2\omega^4}\left[-\tfrac{3}{16} \sin 3\omega t - \tfrac{15}{4}\omega t \cos \omega t + \alpha \sin \omega t\right], \tag{7.117}$$

where α is at the moment an arbitrary, adjustable parameter.

The $p^{(i)}$ are related to the $q^{(i)}$ by the equation

$$p^{(i)} = m\dot{q}^{(i)}, \tag{7.118}$$

and one gets for the energy

$$E = E^{(0)} + \lambda E^{(1)} + \lambda^2 E^{(2)} + \ldots \tag{7.119}$$

with

$$E^{(0)} = \tfrac{1}{2}m\omega^2 A^2, \qquad E^{(1)} = 0, \qquad E^{(2)} = \frac{A^4}{m\omega^2}(\alpha - \tfrac{37}{16}). \tag{7.120}$$

In this particular case this method is not very suitable for two reasons: first of all we see that the solutions of the differential equations for $q^{(1)}$ and $q^{(2)}$ allow a term to be added to them, proportional to the solution of the unperturbed equation (7.112). We have not added this term to the solution for $q^{(1)}$, but we have added it to the solution for $q^{(2)}$. This was done in order to be able to get the correct answer — that is, the answer obtained by either the second method or canonical perturbation theory — for $E^{(1)}$ and $E^{(2)}$. If we put α equal to 11/8, this objective is, indeed, obtained [compare (7.133)]. The second difficulty is the occurrence of a term proportional to $t \cos \omega t$ in $q^{(2)}$. If power series expansions in λ are to have any meaning, it must be possible to choose λ so small that the terms in λ^n ($n \neq 0$) are smaller than the unperturbed terms. This is clearly impossible if an unbounded term such as $t \cos \omega t$ occurs. We shall see in the next section that, indeed, this term does not occur in the expression for $q^{(2)}$ obtained by canonical perturbation theory [see (7.235)]. We must add that these difficulties are peculiar to the harmonic oscillator, since in that case, and in that case only, the unperturbed equation of motion is a homogeneous linear differential equation. Inasfar as the harmonic oscillator equations of motion are possibly the most important ones in physics, the objections raised here are serious ones.

Let us now apply the second method to the anharmonic oscillator. We shall see that this method does not encounter the above-mentioned difficulties, but that, on the other hand, it is a very cumbersome way of getting the perturbed motion. Indeed, the simplest method to obtain the first non-vanishing corrections to either the energy or the q is the canonical perturbation theory, discussed in the next section.

We shall follow here Born's treatment [Vorlesungen über Atommechanik (Springer, Berlin, 1925) p. 76] and use action and angle variables. If E be the energy of the system, we have

$$E = (p^2/2m) + \tfrac{1}{2}m\omega^2 q^2 + \lambda q^3,$$

or

$$p = \sqrt{2m\lambda f(q)}, \tag{7.121}$$

with

$$f(q) = -q^3 - (m\omega^2 q^2/2\lambda) + E/\lambda. \tag{7.122}$$

Let e_1, e_2, and e_3 be the three zeros of $f(q)$, which we shall choose such that $e_1 \to +(2E/m\omega^2)^{\frac{1}{2}}$, $e_2 \to -(2E/m\omega^2)^{\frac{1}{2}}$, and $e_3 \to -\infty$ as $\lambda \to 0$ (see fig. 32). We can write (7.122) now as follows

$$\begin{aligned} f(q) &= (e_1 - q)(q - e_2)(q - e_3) \\ &= -e_3(e_1 - q)(q - e_2)[1 - (q/e_3)]. \end{aligned} \tag{7.123}$$

We shall look for the exact solution of the equations of motion in terms of a power series in λ in the limit as $\lambda \to 0$. In that limit e_3 will tend to infinity as $-\lambda^{-1}$, and taking the square root of equation (7.123), we can

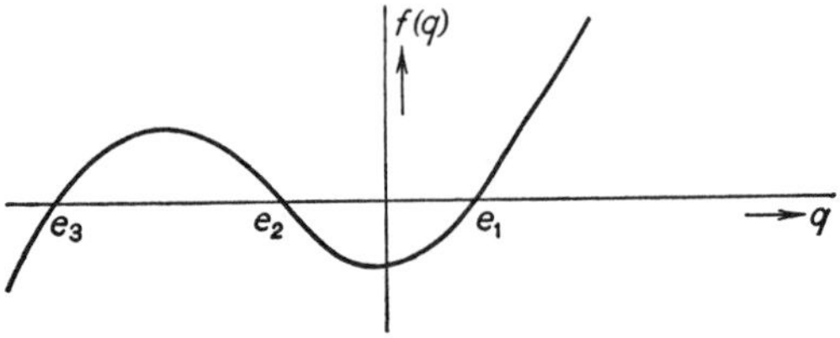

Fig. 32. The function $f(q)$ defined by equation (7.122); e_1, e_2, and e_3 are the roots of $f(q)$.

expand the square root of the last factor in a power series in e_3^{-1},

$$[f(q)]^{\frac{1}{2}} = (-e_3)^{\frac{1}{2}}[(e_1-q)(q-e_2)]^{\frac{1}{2}}[1-(q/2e_3)-(q^2/8e_3^2)+\ldots]. \quad (7.124)$$

From (6.210) and (7.121) we now get for the action variable J

$$J = \oint p\,\mathrm{d}q = (-2m\lambda e_3)^{\frac{1}{2}}[J^{(0)}-(J^{(1)}/2e_3)-(J^{(2)}/8e_3^2)+\ldots], \quad (7.125)$$

where

$$J^{(k)} = \oint [(e_1-q)(q-e_2)]^{\frac{1}{2}} q^k \,\mathrm{d}q. \quad (7.126)$$

If we introduce instead of q a new variable ψ by the equation [compare (6.230)]

$$2q = (e_1+e_2)+(e_1-e_2)\sin\psi, \quad (7.127)$$

ψ varies between $\pi/2$ and $5\pi/2$ as q goes from e_1 to e_2 and back, so that we get for the $J^{(k)}$

$$J^{(k)} = \tfrac{1}{4}(e_1-e_2)^2 \int_0^{2\pi} [\tfrac{1}{2}(e_1+e_2)+\tfrac{1}{2}(e_1-e_2)\sin\psi]^k \cos^2\psi\,\mathrm{d}\psi, \quad (7.128)$$

which leads to the following expressions for $J^{(0)}$, $J^{(1)}$ and $J^{(2)}$,

$$J^{(0)} = \tfrac{1}{4}\pi(e_1-e_2)^2, \qquad J^{(1)} = \tfrac{1}{8}\pi(e_1+e_2)(e_1-e_2)^2,$$
$$J^{(2)} = \tfrac{1}{64}\pi(e_1-e_2)^2[5(e_1+e_2)^2-4e_1e_2]. \quad (7.129)$$

We now write

$$e_{1,2} = \pm(2E/m\omega^2)^{\frac{1}{2}}+\alpha_{1,2}\lambda+\beta_{1,2}\lambda^2+\ldots,$$
$$e_3 = (-m\omega^2/\lambda)(1+\alpha_3\lambda+\beta_3\lambda^2+\ldots), \quad (7.130)$$

where the α_i and β_i are obtained by successively substituting the expressions for e_1, e_2, and e_3 into expression (7.122) for $f(q)$, requiring it to vanish, and thus equating the coefficients of the various powers of λ to zero. The result is

$$\begin{aligned} \alpha_1 = \alpha_2 &= -2E/m^2\omega^4, \qquad \alpha_3 = 0, \\ \beta_1 = -\beta_2 &= (5E/m^3\omega^7)(2E/m)^{\frac{1}{2}}; \qquad \beta_3 = -8E/m^3\omega^6. \end{aligned} \tag{7.131}$$

Using (7.123) and (7.127) to (7.129) we find

$$J = \frac{2\pi E}{\omega}\left[1+\frac{15\lambda^2 E}{4m^3\omega^6}+\ldots\right], \tag{7.132}$$

or

$$E = \nu J - \frac{15\lambda^2 J^2}{16\pi^2 m^3\omega^4}+\ldots, \qquad \nu = \omega/2\pi. \tag{7.133}$$

To find q in terms of the angle variable w, we use (6.208), or,

$$w = \int \frac{\partial p}{\partial J}\,\mathrm{d}q = \left(\frac{m}{2\lambda}\right)^{\frac{1}{2}} \frac{\mathrm{d}E}{\mathrm{d}J}\int \frac{\mathrm{d}q}{\sqrt{f(q)}}, \tag{7.134}$$

where we have used (7.121) and (7.122), and the fact that in (7.122) only E depends on J.

Expanding $[f(q)]^{-\frac{1}{2}}$ in a way similar to the expansion (7.124), and using the substitution (7.127) we find by straightforward, tedious calculations

$$2\pi w = \psi + \lambda(J/\pi m^3\omega^5)^{\frac{1}{2}} \cos\psi + \ldots \tag{7.135}$$

Combining (7.127), (7.130), (7.131), (7.133), and (7.135) we find finally for q the expression

$$q = A\sin 2\pi w - \frac{\lambda A^2}{2m\omega^2}(3+\cos 4\pi w)+\ldots, \qquad A = \left(\frac{J}{\pi m\omega}\right)^{\frac{1}{2}}. \tag{7.136}$$

2. CANONICAL PERTURBATION THEORY

In this section we shall develop a systematic method for obtaining solutions of the Hamiltonian equations of motion. We have for the Hamiltonian

$$H = H_0 + \lambda H_1 + \lambda^2 H_2 + \ldots \tag{7.201}$$

Although usually there is only one extra term on the right hand side of (7.201), as in all cases considered in this chapter, we have allowed for the occurrence of higher order terms by including a term $\lambda^2 H_2$.

We shall assume that our problem is one which can be solved by using action and angle variables, that is, one corresponding to a separable Hamilton-Jacobi equation (see § 6.2). To simplify the equations we shall restrict our discussion for the present to one-dimensional systems. We shall discuss at the end of this section the extension of the theory to more-dimensional problems. The action and angle variables corresponding to the unperturbed system with the Hamiltonian H_0 will be denoted by J_0 and w_0. The equations of motion are solved by the Hamilton-Jacobi method, and we look for a Hamilton-Jacobi function in the form of a power series in λ,

$$S = S_0 + \lambda S_1 + \lambda^2 S_2 + \ldots. \tag{7.202}$$

This Hamilton-Jacobi function will generate a transformation from the unperturbed w_0 and J_0 to the perturbed w and J through the equations

$$J_0 = \frac{\partial S}{\partial w_0}, \qquad w = \frac{\partial S}{\partial J}, \qquad S = S(w_0, J). \tag{7.203}$$

The original coordinate q is a known function of w_0 and J_0, as we assumed the unperturbed problem to have been solved by introducing action and angle variables. As $\lambda \to 0$, we go over to the unperturbed problem so that S_0 should correspond to the identity transformation; it will thus be of the form [compare (5.224)]

$$S_0 = w_0 J. \tag{7.204}$$

Equations (7.203) will lead to

$$J_0 = J + \lambda \frac{\partial S_1}{\partial w_0} + \lambda^2 \frac{\partial S_2}{\partial w_0} + \ldots, \qquad w = w_0 + \lambda \frac{\partial S_1}{\partial J} + \lambda^2 \frac{\partial S_2}{\partial J} + \ldots. \tag{7.205}$$

The $S_1, S_2, \ldots$ are functions of w_0 and J, but we can use (7.205) to find J_0 and w_0 as power series in λ with each term depending on w and J:

$$J_0 = J + \lambda \left(\frac{\partial S_1}{\partial w_0}\right)_{w_0 = w} + \lambda^2 \left[\left(\frac{\partial S_2}{\partial w_0}\right)_{w_0 = w} - \left(\frac{\partial^2 S_1}{\partial w_0^2}\frac{\partial S_1}{\partial J}\right)_{w_0 = w}\right] + \ldots; \tag{7.206}$$

$$w_0 = w - \lambda \left(\frac{\partial S_1}{\partial J}\right)_{w_0 = w} + \lambda^2 \left[-\left(\frac{\partial S_2}{\partial J}\right)_{w_0 = w} + \left(\frac{\partial^2 S_1}{\partial J \partial w_0}\frac{\partial S_1}{\partial J}\right)_{w_0 = w}\right] + \ldots. \tag{7.207}$$

Originally the $H_0, H_1, H_2, \ldots$ were functions of p and q, but after the first Hamilton-Jacobi transformation, which introduces w_0 and J_0 and at the same time solves the unperturbed problem, we have

$$H = H_0(J_0)+\lambda H_1(w_0, J_0)+\lambda^2 H_2(w_0, J_0)+ \dots . \tag{7.208}$$

The new Hamilton-Jacobi equation to be solved for S will be

$$H_0\left(\frac{\partial S}{\partial w_0}\right)+\lambda H_1\left(w_0, \frac{\partial S}{\partial w_0}\right)+\lambda^2 H_2\left(w_0, \frac{\partial S}{\partial w_0}\right)+ \dots = E(J). \tag{7.209}$$

Using (7.202) for S, expanding the energy $E(J)$ as a power series in λ,

$$E(J) = E_0(J)+\lambda E_1(J)+\lambda^2 E_2(J)+ \dots, \tag{7.210}$$

and using (7.204) and (7.205) we get from (7.209)

$$\begin{aligned} H_0(J)+\lambda \frac{\partial H_0}{\partial J_0}\frac{\partial S_1}{\partial w_0}+\lambda^2\frac{\partial H_0}{\partial J_0}\frac{\partial S_2}{\partial w_0}+\tfrac{1}{2}\lambda^2\frac{\partial^2 H_0}{\partial J_0^2}\left(\frac{\partial S_1}{\partial w_0}\right)^2+\dots \\ +\lambda H_1(w_0, J)+\lambda^2\frac{\partial H_1}{\partial J_0}\frac{\partial S_1}{\partial w_0}+\dots+\lambda^2 H_2(w_0, J)+\dots \\ = E_0(J)+\lambda E_1(J)+\lambda^2 E_2(J)+\dots . \end{aligned} \tag{7.211}$$

Combining all terms corresponding to the same power of λ, we get

$$H_0(J) = E_0(J), \tag{7.212}$$

$$\frac{\partial H_0}{\partial J_0}\frac{\partial S_1}{\partial w_0}+H_1(w_0, J) = E_1(J), \tag{7.213}$$

$$\frac{\partial H_0}{\partial J_0}\frac{\partial S_2}{\partial w_0}+\frac{1}{2}\frac{\partial^2 H_0}{\partial J_0^2}\left(\frac{\partial S_1}{\partial w_0}\right)^2+\frac{\partial H_1}{\partial J_0}\frac{\partial S_1}{\partial w_0}+H_2(w_0, J) = E_2(J), \dots . \tag{7.214}$$

Equation (7.213) determines S_1 and E_1, equation (7.214) S_2 and $E_2, \dots$ To see how this is done in practice, we remind ourselves that the p and q are periodic functions of w_0 with period 1 (see § 6.2). This will, of course, still be the case, even though w_0 is no longer a linear function of the time. Since p and q are periodic in w_0, the same will be true for $H_1, H_2, \dots$. We can therefore expand these quantities in Fourier series

$$H_1 = \sum_n H_n^{(1)}(J_0) \exp(2\pi i n w_0), \tag{7.215}$$

$$H_2 = \sum_n H_n^{(2)}(J_0) \exp(2\pi i n w_0), \dots . \tag{7.216}$$

It is now convenient to expand $S_1, S_2, \dots$ also in Fourier series:

$$S_1 = \sum_n S_n^{(1)}(J) \exp(2\pi i n w_0), \tag{7.217}$$

$$S_2 = \sum_n S_n^{(2)}(J) \exp(2\pi i n w_0), \dots . \tag{7.218}$$

From (7.217) and (7.218) it follows that in the Fourier series for $\partial S_1/\partial w_0$ and $\partial S_2/\partial w_0$ there do not occur any terms with $n = 0$, and that thus the averages of $\partial S_1/\partial w_0$ and $\partial S_2/\partial w_0$ over one period will be equal to zero. On the other hand, if one evaluates the averages of $H_1, H_2, \dots$ or their derivatives with respect to J_0 over one period, one finds just the term corresponding to $n = 0$.

Taking therefore the average of equations (7.213) and (7.214) over one period — we denote these averages by a bar — we get

$$E_1(J) = \overline{H}_1, \tag{7.219}$$

$$E_2(J) = \overline{H}_2 + \overline{\frac{1}{2}\frac{\partial^2 H_0}{\partial J_0^2}\left(\frac{\partial S_1}{\partial w_0}\right)^2} + \overline{\frac{\partial H_1}{\partial J_0}\frac{\partial S_1}{\partial w_0}}, \dots . \tag{7.220}$$

We note that although $\overline{\partial S_1/\partial w_0} = 0$, the last two terms on the right hand side of equation (7.220) do not necessarily vanish. Equations (7.219) and (7.220) determine E_1 and E_2.

If we write

$$\tilde{G} = G - \overline{G} \tag{7.221}$$

for any function G of w_0, that is, if we denote the purely periodic part of a function of w_0 by a tilde, we get from (7.213), (7.214), (7.219), and (7.220) the following equations from which we can solve for S_1 and S_2:

$$\frac{\partial S_1}{\partial w_0} = \frac{-\widetilde{H_1(J)}}{\nu}, \tag{7.222}$$

$$\frac{\partial S_2}{\partial w_0} = \frac{1}{\nu}\left[-\tilde{H}_2 - \widetilde{\frac{1}{2}\frac{{}^2 H_0}{\partial J_0^2}\left(\frac{\partial S_1}{\partial w_0}\right)^2} - \widetilde{\frac{\partial H_1}{\partial J_0}\frac{\partial S_1}{\partial w_0}}\right], \tag{7.223}$$

where we have put [compare (7.212) and (6.211)]

$$\frac{\partial H_0}{\partial J_0} = \nu. \tag{7.224}$$

One can express S_1 and S_2 in terms of the Fourier coefficients $H_n^{(1)}$ and $H_n^{(2)}$ of equations (7.215) and (7.216), but we do not wish to do this at this juncture.

Before discussing more-dimensional systems, we shall apply the theory developed here to the case discussed in the preceding section. When dealing with the anharmonic oscillator, we must first of all remind ourselves of the Hamilton-Jacobi solution of the unperturbed problem. From (7.136) we see that expressed in terms of w_0 and J_0 the coordinate q is given by the equation

$$q = (J_0/\pi m\omega)^{\frac{1}{2}} \sin 2\pi w_0 . \tag{7.225}$$

In terms of w_0 and J_0 the total Hamiltonian is now

$$H = H_0 + \lambda H_1 ,$$

with

$$H_0 = \nu J_0 , \qquad H_1 = (J_0/\pi m\omega)^{\frac{3}{2}} \sin^3 2\pi w_0 , \tag{7.226}$$

where ν is the quantity introduced in (7.133), that is, $\nu = \omega/2\pi$ [compare also (7.224)].

From (7.219) and (7.222) we get

$$E_1 = 0, \tag{7.227}$$

$$\frac{\partial S_1}{\partial w_0} = -\frac{1}{\nu}\left(\frac{J}{\pi m\omega}\right)^{\frac{3}{2}} \sin^3 2\pi w_0 , \tag{7.228}$$

or

$$S_1 = \frac{1}{2\pi\nu}\left(\frac{J}{\pi m\omega}\right)^{\frac{3}{2}} [\tfrac{3}{4}\cos 2\pi w_0 - \tfrac{1}{12}\cos 6\pi w_0], \tag{7.229}$$

from which follows up to terms linear in λ

$$q = \left(\frac{J}{\pi m\omega}\right)^{\frac{1}{2}} \sin 2\pi w - \frac{\lambda}{2m\omega^2}\frac{J}{\pi m\omega}(3+\cos 4\pi w), \tag{7.230}$$

in accordance with equation (7.136).

From (7.220) we now get

$$E_2 = -\frac{15J^2}{16\pi^2 m^3\omega^4}, \tag{7.231}$$

in accordance with equation (7.133).

We can use (7.223) to find $\partial S_2/\partial w_0$ and thus S_2. The result is

$$S_2 = \frac{3J^2}{4\pi^3 m^3\omega^4}[-\tfrac{15}{32}\sin 4\pi w_0 + \tfrac{3}{32}\sin 8\pi w_0 - \tfrac{1}{96}\sin 12\pi w_0]. \tag{7.232}$$

After a tedious calculation one finds up to terms in λ^2 for J_0, w_0, and q in terms of J and w

$$J_0 = J + \frac{\lambda}{\nu}\left(\frac{J}{\pi m\omega}\right)^{\frac{3}{2}}(\tfrac{1}{4}\sin 6\pi w - \tfrac{3}{4}\sin 2\pi w)$$

$$+ \frac{3\lambda^2 J^2}{2\nu^2(\pi m\omega)^3}(-\tfrac{1}{8}\cos 8\pi w - \tfrac{1}{2}\cos 4\pi w + \tfrac{5}{16}) + \ldots, \tag{7.233}$$

$$w_0 = w + \frac{3\lambda}{2\omega J}\left(\frac{J}{\pi m\omega}\right)^{\frac{3}{2}}(\tfrac{1}{12}\cos 6\pi w - \tfrac{3}{4}\cos 2\pi w)$$

$$+ \frac{3\lambda^2 J}{2\nu\omega(\pi m\omega)^3}(\tfrac{9}{64}\sin 4\pi w + \tfrac{3}{32}\sin 8\pi w - \tfrac{1}{192}\sin 12\pi w) + \ldots, \tag{7.234}$$

$$q = \left(\frac{J}{\pi m\omega}\right)^{\frac{1}{2}}\sin 2\pi w - \frac{\lambda}{2m\omega}\frac{J}{\pi m\omega}(3 + \cos 4\pi w)$$

$$+ \frac{\lambda^2}{m^2\omega^4}\left(\frac{J}{\pi m\omega}\right)^{\frac{3}{2}}(-\tfrac{3}{16}\sin 6\pi w + \tfrac{11}{8}\sin 2\pi w) + \ldots \tag{7.235}$$

We note that, apart from the unbounded term, $q^{(2)}$ given by (7.117) with $\alpha = \frac{11}{8}$ agrees with the λ^2 coefficient in (7.235). We also note that canonical perturbation theory is very fast, if one wants to obtain the change in energy, though more cumbersome, if one needs the change in the coordinates. In both respects, it is quicker than the second method of the preceding section. We must remind ourselves that the final expression for the energy is a function of J only so that w will be a linear function of the time.

To conclude this section, we shall discuss briefly the case of more-dimensional problems, still restricting ourselves to the case where the unperturbed problem can be solved by separation of variables together with the introduction of action and angle variables; in other words we consider multiply periodic systems. Let $w_k^{(0)}$ and $J_k^{(0)}$ ($k = 1, \ldots, s$; s: number of degrees of freedom) be the action and angle variables of the unperturbed problem. The q_k will therefore all be periodic functions of the $w_k^{(0)}$, and so will the $H_1, H_2, \ldots$ in (7.201). We shall now throughout use the Fourier expansions which we mentioned at the beginning of this section, and we write by analogy with (7.215) and (7.216)

$$H_1 = \sum_{n_1, n_2, \ldots} H^{(1)}_{n_1, n_2, \ldots}(J_1^{(0)}, \ldots) \exp 2\pi i \sum_k n_k w_k^{(0)}, \tag{7.236}$$

$$H_2 = \sum_{n_1, n_2, \ldots} H^{(2)}_{n_1, n_2, \ldots}(J_1^{(0)}, \ldots) \exp 2\pi i \sum_k n_k w_k^{(0)}, \tag{7.237}$$

where we have now multiple Fourier series.

We now look for $S_1, S_2, \ldots$ in the form of Fourier series,

$$S_1 = \sum_{n_1, n_2, \ldots} S^{(1)}_{n_1, n_2, \ldots}(J_1, \ldots) \exp 2\pi i \sum_k n_k w_k^{(0)}, \tag{7.238}$$

$$S_2 = \sum_{n_1, n_2, \ldots} S^{(2)}_{n_1, n_2, \ldots}(J_1, \ldots) \exp 2\pi i \sum_k n_k w_k^{(0)}. \tag{7.239}$$

Instead of (7.213) we have now

$$\sum_k \frac{\partial H_0}{\partial J_k^{(0)}} \frac{\partial S_1}{\partial w_k^{(0)}} + H_1(w_k^{(0)}, J_k) = E_1(J_k), \tag{7.240}$$

and a similar equation for E_2. Indicating now by a bar across an average over all $w_k^{(0)}$, that is,

$$\bar{G} = \int_0^1 \cdots \int \mathrm{d}w_1^{(0)} \ldots \mathrm{d}w_s^{(0)} G, \tag{7.241}$$

we find from (7.240)

$$E_1 = \bar{H}_1 = H^{(1)}_{00\ldots}. \tag{7.242}$$

We can then solve (7.240) for S_1, and we find

$$S^{(1)}_{n_1, n_2, \ldots} = \frac{-H^{(1)}_{n_1, n_2, \ldots}}{2\pi i \sum_k n_k \nu_k^{(0)}}, \tag{7.243}$$

where

$$\nu_k^{(0)} = \frac{\partial H_0}{\partial J_k^{(0)}}. \tag{7.244}$$

The higher order terms in S can be obtained in a similar way. We see from (7.243) that trouble arises, if the $\nu_k^{(0)}$ are such that one can choose a set $n_1, \ldots, n_s$ which are not all equal to zero such that

$$\sum_k n_k \nu_k^{(0)} = 0. \tag{7.245}$$

Such a case is called a *degenerate* case. If this case is excluded one can still find such combinations of the n_k that $\sum n_k \nu_k^{(0)}$ becomes as small as one likes. This has as a result that for practically all perturbations H_1 the Fourier series (7.238) and (7.239) diverge, as was shewn by Poincaré. These series are, however, semi-convergent, that is, suitably truncated these series can with great accuracy predict the behaviour of the system for very long, though not arbitrarily long, periods,

Having found $S^{(1)}$ (apart from $S^{(1)}_{00\ldots}$ which we shall put equal to zero) and thus from equations similar to (7.206) and (7.207) the new action and angle variables, we can then express the coordinates in terms of the new variables and thus find the behaviour of the system.

We have emphasised a moment ago that canonical perturbation theory for the case of more than one degree of freedom leads to divergent series. It is sometimes convenient — and we shall see an example in the next section — to use the old $w_k^{(0)}$ and $J_k^{(0)}$, which, of course, remain canonically conjugate variables for the perturbed system since they were obtained from the p_k and q_k by a canonical transformation, to solve the equations of motion. This is especially the case when one is dealing with a degenerate system. The simplest case of degeneracy is the one we met with in Chapter 6 where some of the $\nu_k^{(0)}$ are the same. In the Kepler problem we found that $\nu_1 = \nu_2 = \nu_3$. In that case, we can use instead of the J's given by (6.224) to (6.226) any linear combination of them, and especially 2π times the α_1, α_2, and α_3 which were introduced in §6.1. If we denote 2π times these α_i by $J_1^{(0)}, J_2^{(0)}$, and $J_3^{(0)}$, and the canonically conjugate variables by $w_1^{(0)}$, $w_2^{(0)}$, and $w_3^{(0)}$, we have an unperturbed system for which

$$H_0 = H_0(J_1^{(0)}), \tag{7.246}$$

and for which we have the (unperturbed) equations of motion

$$\dot{J}_1^{(0)} = 0, \quad \dot{J}_2^{(0)} = 0, \quad \dot{J}_3^{(0)} = 0, \quad \dot{w}_1^{(0)} = \nu_1^{(0)}, \quad \dot{w}_2^{(0)} = 0, \quad \dot{w}_3^{(0)} = 0, \tag{7.247}$$

where $\nu_1^{(0)}$ is given by (7.244).

If we now consider the perturbed Hamiltonian

$$H = H_0 + \lambda H_1, \tag{7.248}$$

we get the equations of motion

$$\begin{aligned}
\dot{w}_1^{(0)} &= \nu_1^{(0)} + \lambda \frac{\partial H_1}{\partial J_1^{(0)}}, & \dot{J}_1^{(0)} &= -\lambda \frac{\partial H_1}{\partial w_1^{(0)}}, \\
\dot{w}_2^{(0)} &= \lambda \frac{\partial H_1}{\partial J_2^{(0)}}, & \dot{J}_2^{(0)} &= -\lambda \frac{\partial H_1}{\partial w_2^{(0)}}, \\
\dot{w}_3^{(0)} &= \lambda \frac{\partial H_1}{\partial J_3^{(0)}}, & \dot{J}_3^{(0)} &= -\lambda \frac{\partial H_1}{\partial w_3^{(0)}}.
\end{aligned} \tag{7.249}$$

We once again introduce Fourier series, and we write for H_1

$$H_1 = \sum_{n_1, n_2, n_3 = -\infty}^{+\infty} H^{(1)}_{n_1, n_2, n_3}(J_i^{(0)}) \exp 2\pi i \sum_{k=1}^{3} n_k w_k^{(0)}. \tag{7.250}$$

Indicating by a prime the sum over n_1 which does not include the term with $n_1 = 0$, we write (7.250) in the form

$$H_1 = F(J_i^{(0)}; w_2^{(0)}, w_3^{(0)}) + G(J_i^{(0)}; w_1^{(0)}, w_2^{(0)}, w_3^{(0)}), \tag{7.251}$$

where

$$\begin{aligned}
F &= \sum_{n_2, n_3} H^{(1)}_{0 n_2 n_3} \exp 2\pi i (n_2 w_2^{(0)} + n_3 w_3^{(0)}), \\
G &= \sum_{n_1}{}' \sum_{n_2, n_3} H^{(1)}_{n_1 n_2 n_3} \exp 2\pi i (n_1 w_1^{(0)} + n_2 w_2^{(0)} + n_3 w_3^{(0)}).
\end{aligned} \tag{7.252}$$

We now proceed to solve the equations of motion by the method of successive approximations. This means that we can substitute into H_1 the solutions of the unperturbed equations of motion (7.247). These solutions are that $w_2^{(0)}$, $w_3^{(0)}$, $J_1^{(0)}$, $J_2^{(0)}$, and $J_3^{(0)}$ are all constants while $w_1^{(0)}$ is a linear function of the time. We see thus that in (7.251) F does not contain the time while G is a periodic function of the time. In fact F is just the time average of H_1. Let us now consider the perturbed equations of motion, for instance, the equation for $J_2^{(0)}$. This equation can be written as

$$\dot{J}_2^{(0)} = -\lambda \frac{\partial F}{\partial w_2^{(0)}} - \lambda \frac{\partial G}{\partial w_2^{(0)}}, \tag{7.253}$$

and integrating we find

$$J_2^{(0)}(t) = J_2^{(0)}(0) - \lambda \frac{\partial F}{\partial w_2^{(0)}} t - \sum_{n_1}{}' \sum_{n_2, n_3} \frac{\lambda H^{(1)}_{n_1 n_2 n_3} n_2}{n_1 \nu_1^{(0)}} \exp 2\pi i \sum_k n_k w_k^{(0)}. \tag{7.254}$$

We find two corrections to $J_2^{(0)}$: the first one, stemming from G, is a periodical one and will remain bounded, but the second one, coming from F, is a *secular* term which increases linearly with time. The solution (7.254) is therefore clearly valid only for a sufficiently small time interval.

Similar situations arise for the other variables, except $J_1^{(0)}$ as $\partial F/\partial w_1^{(0)} = 0$. We do not wish to discuss here how one should treat systems where both secular and periodic perturbations occur, but we must point out that in contradistinction to quantum mechanics where one can easily get rid of secular effects, this is not the case in classical mechanics. For a further discussion of these problems we refer to the literature given in the bibliography.

3. ZEEMAN AND STARK EFFECT OF THE HYDROGEN ATOM

Let us consider a one-particle system the Hamiltonian of which is spherically symmetric. In that case we know that if we introduce spherical polars r, θ, and φ, the angle φ will be a cyclic variable. The corresponding momentum p_φ can then be introduced as one of the α_i (see § 6.1) and the corresponding action variable J_φ is given by the equation [compare (6.224)]

$$J_\varphi = 2\pi p_\varphi . \tag{7.301}$$

The corresponding angle variable w_φ can be chosen to be the length of the ascending node (see § 6.1), divided by 2π.

Let us now assume that a uniform magnetic field $\boldsymbol{B}$ is applied. As our unperturbed system was spherically symmetric, we can choose our polar axis arbitrarily, and we shall now fix it along the direction of the magnetic field. We have seen earlier that the influence of a magnetic field can be taken into account by changing in the kinetic energy the term $\boldsymbol{p}^2$ to $(\boldsymbol{p}-e\boldsymbol{A})^2$, where $\boldsymbol{A}$ is the vector potential [see (5.355)]. We can choose $\boldsymbol{A}$ to be of the form

$$\boldsymbol{A} = \tfrac{1}{2}[\boldsymbol{B} \wedge \boldsymbol{x}], \tag{7.302}$$

and assuming the magnetic field to be weak we have for the perturbed Hamiltonian

$$H = H_0 + H_1 , \tag{7.303}$$

where

$$\begin{aligned} H_1 &= \frac{(\boldsymbol{p}-e\boldsymbol{A})^2}{2m} - \frac{\boldsymbol{p}^2}{2m} \doteqdot -\frac{e(\boldsymbol{A}\cdot\boldsymbol{p})}{m} \\ &= -\frac{e}{2m}(\boldsymbol{p}\cdot[\boldsymbol{B}\wedge\boldsymbol{x}]) = -\frac{e}{2m}(\boldsymbol{B}\cdot[\boldsymbol{x}\wedge\boldsymbol{p}]) = -\frac{e}{2m}Bp_\varphi . \end{aligned} \tag{7.304}$$

The field strength B plays the role here of the expansion parameter λ of the previous sections.

As H_1 contains only J_φ, the only variable which is involved in the first approximation will be w_φ. From the canonical equations of motion we get

$$\dot{w}_\varphi = \frac{\partial H}{\partial J_\varphi} = \frac{\partial H_0}{\partial J_\varphi} + \frac{\partial H_1}{\partial J_\varphi} = \frac{\partial H_1}{\partial J_\varphi} = -\frac{eB}{4\pi mc}. \tag{7.305}$$

We see thus that the orbital plane rotates around the direction of the magnetic field with an angular velocity $eB/2mc$, the so-called Larmor frequency, and this rotation of the orbital plane is called the Larmor precession.

The case of a weak uniform electric field is more complicated. One way to treat this problem is by the second method described in the first section of this chapter (see problem 5 (ch. 6), problem section). We shall not consider here the whole problem, but only the secular effects of a uniform electric field. We shall do this first of all by elementary considerations — which can be extended to a discussion of crossed electric and magnetic fields (see problem 3 in the problem section) — and then by applying the theory of secular perturbations discussed briefly at the end of the previous section. In all cases, we shall only consider the case of the hydrogen atom and we shall use the results obtained in the preceding chapter.

We shall again choose the z-axis to be in the direction of the field so that the perturbing Hamiltonian is given by the equation

$$H_1 = e\mathscr{E}z, \tag{7.306}$$

where $\mathscr{E}$ is the field strength.

We shall use as our J's and w's the α_i and β_i of § 6.2, and we remind ourselves of the relationships between the semi-major axis a, the total angular momentum M, the eccentricity ε, and the inclination of the orbital plane, i, and α_1, α_2, and α_3, on the one hand, and between the time, the length of the pericentre, and the length of the ascending node and β_1, β_2, and β_3, on the other hand. These relationships were all derived in § 6.1 and will be used here.

We saw in the preceding section that the secular perturbation are determined by the time average $\overline{H}_1$ of the perturbing energy. We must therefore evaluate the time average of the position of the electron, which at the same time gives us the values of the 'centre of charge' of the hydrogen atom. We shall see that it does not coincide with the origin for an elliptic orbit. The atom will thus act as an electric dipole, and we should expect secular effects to arise from the action of an electric field.

The reason why the centre of charge is different from the centre of mass (which is the origin, or the focus of the ellipse) and from the centre of the ellipse is that the electron is moving faster near the pericentre than near the apocentre and therefore spends more time in those parts of the orbit which are near the apocentre.

To evaluate $\bar{z}$ we first of all calculate the average values of the rectangular coordinates ξ and η introduced in fig. 28 and given by equations (6.153). Using the fact that β_1 is essentially the time [see (6.155)] that β_1 is related to u by (6.156), and that the period of u is 2π, we have

$$\begin{aligned} \bar{\xi} &= \frac{1}{2\pi}\int_0^{2\pi} a(\cos u - \varepsilon)(1-\varepsilon\cos u)\mathrm{d}u = -\tfrac{3}{2}a\varepsilon, \\ \bar{\eta} &= \frac{1}{2\pi}\int_0^{2\pi} a(1-\varepsilon^2)^{\frac{1}{2}}\sin u(1-\varepsilon\cos u)\mathrm{d}u = 0. \end{aligned} \tag{7.307}$$

If we now use fig. 26 and the fact that it follows from equations (7.307) that the 'centre of charge' is situated along OP we find

$$\bar{z} = \bar{\xi}\sin\beta_2\sin i = -\tfrac{3}{2}\varepsilon a\sin\beta_2\sin i. \tag{7.308}$$

Before using this result, we shall use a more elementary method to discuss the influence of the electric field $\mathscr{E}$. The field will exert a torque $-e[\boldsymbol{x}\wedge\mathscr{E}]$ on the orbit where $\boldsymbol{x}$ is the position of the electron. If $\boldsymbol{n}$ is a unit vector along the major axis in the apocentre direction, we have from equation (7.307) for the time average of $\boldsymbol{x}$ the value $\tfrac{3}{2}\varepsilon a\boldsymbol{n}$, and thus for the time average of the torque $-\tfrac{3}{2}\varepsilon ae[\boldsymbol{n}\wedge\mathscr{E}]$. We shall now consider two cases (i) $\mathscr{E}$ perpendicular to the orbital plane, and (ii) $\mathscr{E}$ in the orbital plane, making an angle ψ with the major axis.

Case (i). The equation of motion we are interested in gives the rate of change of the angular momentum $\boldsymbol{M}$:

$$\frac{\mathrm{d}\boldsymbol{M}}{\mathrm{d}t} = \text{torque} = -e\,\overline{[\boldsymbol{x}\wedge\mathscr{E}]},$$

or

$$\frac{\mathrm{d}\boldsymbol{M}}{\mathrm{d}t} = \tfrac{3}{2}e\varepsilon a[\boldsymbol{n}\wedge\mathscr{E}]. \tag{7.309}$$

The vector $\boldsymbol{M}$ is perpendicular to the orbital plane, while $[\boldsymbol{n} \wedge \mathscr{E}]$ is a vector in the orbital plane along the minor axis. We see thus that the orbital plane will rotate around the major axis. The rate of rotation r will be given by the equation

$$r = \tfrac{3}{2} e\varepsilon a \mathscr{E}/M, \tag{7.310}$$

Case (ii). The equation of motion is now

$$\frac{\mathrm{d}M}{\mathrm{d}t} = \tfrac{3}{2} e\varepsilon a \mathscr{E} \sin \psi, \tag{7.311}$$

as now $\boldsymbol{M}$ and $[\boldsymbol{n} \wedge \mathscr{E}]$ are parallel.

The expression $\bar{z}$ is independent of time, by definition, and this can be written as

$$\tfrac{3}{2} \varepsilon a \cos \psi = \text{constant}. \tag{7.312}$$

We saw in the preceding section that α_1 is not subject to secular perturbations. This means that a is constant. Equation (7.312) gives us thus a relation between the rate of change of ε and of ψ, that is, between the rate of change of the eccentricity and of the orientation of the orbit. From equations (7.311), (7.312), and the relation (6.150) between M ($= \alpha_2$) and ε (taking into account that α_1 is a constant) we get for the rates of change of ε and of ψ the equations

$$\begin{aligned} \frac{\mathrm{d}\varepsilon}{\mathrm{d}t} &= -\frac{3e\mathscr{E}a}{2\alpha_1}(1-\varepsilon^2)^{\frac{1}{2}} \sin \psi, \\ \frac{\mathrm{d}\psi}{\mathrm{d}t} &= -\frac{3e\mathscr{E}a}{2\alpha_1}\frac{(1-\varepsilon^2)^{\frac{1}{2}}}{\varepsilon} \cos \psi. \end{aligned} \tag{7.313}$$

The orbit will be a rosette in the orbital plane.

Another way to treat this problem is to go back to the equations of motion (7.249). Using (6.147) and (6.150) we see that $\bar{z}$ and thus $\bar{H}_1$ is given by the expression

$$\bar{H}_1 = -\frac{3e\mathscr{E}\alpha_1}{2R\alpha_2}[(\alpha_1^2-\alpha_2^2)(\alpha_2^2-\alpha_3^2)]^{\frac{1}{2}} \sin \beta_2 . \tag{7.314}$$

We see that α_1 and α_3 are not changed. If we substitute this expression for $\bar{H}_1$ into the time average of equations (7.249), we get the equations of motion for the secular changes in α_2, β_1, β_2 and β_3. From these equations we can obtain the equation of motion for the x- and y-coordinates of the centre of charge. In terms of β_2, β_3 and i one finds for the x- and y-coordinates, x and y,

$$\begin{aligned} x &= -\tfrac{3}{2}a\varepsilon(\cos\beta_2\cos\beta_3 - \sin\beta_2\sin\beta_3\cos i), \\ y &= -\tfrac{3}{2}a\varepsilon(\cos\beta_2\sin\beta_3 + \sin\beta_2\cos\beta_3\cos i). \end{aligned} \tag{7.315}$$

Using the fact that α_1 and a are constant and the relations $\varepsilon^2 = 1-(\alpha_2^2/\alpha_1^2)$,

$$\dot{\beta}_2 = \frac{\partial \overline{H}_1}{\partial \alpha_2}, \qquad \dot{\beta}_3 = \frac{\partial \overline{H}_1}{\partial \alpha_3}, \qquad \dot{\alpha}_2 = -\frac{\partial \overline{H}_1}{\partial \beta_2}, \tag{7.316}$$

we get after some tedious calculations the following equations of motion for x and y:

$$\ddot{x} = -\left(\frac{3ae\mathscr{E}}{2\alpha_1}\right)^2 x, \qquad \ddot{y} = -\left(\frac{3ae\mathscr{E}}{2\alpha_1}\right)^2 y, \tag{7.317}$$

and we see that the centre of charge performs a harmonic oscillation with angular frequency ω given by the equation

$$\omega = \frac{6\pi\varepsilon_0\alpha_1}{Zme}\,\mathscr{E}, \tag{7.318}$$

where we have used (6.150) and (6.144) for a.

In conclusion we must mention that we have discussed only a few of the many aspects of perturbation theory. The subject of periodic perturbations, for instance, which is of such great importance in celestial mechanics has been completely neglected.

CHAPTER 8

CONTINUOUS SYSTEMS

In this chapter we discuss how one can treat the equations of motion of continuous systems in the same way as the systems discussed in the preceding chapters. The method used here for obtaining the canonical equations of motion which describe such continuous systems is to introduce the Fourier components of the variables $Q(x)$ describing the system. We then discuss the alterations to be made in the Lagrangian and the Hamiltonian formalism to cover the case of continuous systems. In the second section of this chapter we apply the theory developed in the first section to the case of sound waves and to the case of the electromagnetic field.

1. THE LAGRANGIAN AND HAMILTONIAN FORMALISM FOR CONTINUA

In the previous chapters we have discussed the Lagrangian formalism for systems of point particles and for rigid bodies, and the Hamiltonian formalism for systems of point particles. We have mentioned as one of the advantages of the Hamiltonian formalism that it enables us easily to make the transition to quantum mechanics. The systems discussed so far were all described by a finite number of variables. There are many physical syster which must be described by an infinite number of variables. This usually occurs when instead of variables q_k with $k = 1, \ldots, s$ we have one or more sets of variables $Q(x)$; these $Q(x)$ are functions of a continuous variable, x, just as q_k can be considered to be a function of the discrete variable k. This happens in two different kinds of circumstances. First of all there is the case of continuous media such as gases or liquids, and secondly there is the case of fields. We shall consider instances of both of these cases. It is of interest to see whether such systems can be described by a Lagrangian or Hamiltonian formalism since such a formalism could be used as a basis for quantisation. There are various methods to arrive at a formalism for continuous systems. One which is often used is to consider first a case such as an elastic rod, to treat the continuous system as the limit of a system of point particles, and then to generalise the results obtained for this particular case to the general case. A second method is to assume as a starting point a suitably generalised variational principle. A third method, and the one which we shall adopt as a starting point, is to use instead of the $Q(x)$ their Fourier coefficients Q_k as generalised variables. The advantages of this method are two-fold. First of all we are now dealing with a function of a discrete argument k — at any rate as long as we assume the system to be enclosed within a finite, albeit very large, volume — instead of with a function of a continuous argument, x. Secondly, the theory in its canonical form is more easily quantised and the Fourier coefficients themselves can often be introduced as Jordan-Klein or Jordan-Wigner creation and annihilation operators. The best example for this treatment is probably the electromagnetic field. We shall, however, defer a discussion of that case to the next section. There are some difficulties which are peculiar to the electromagnetic field, connected with the Lorentz gauge condition, and we shall for that reason use as the basis of our treatment the case of longitudinal elastic waves in a one-dimensional continuous medium. This example should illustrate the basic ideas.

The problem with which we are confronted is the following one. We have a set of equations of motion — the Maxwell equations in the electromagnetic case, the wave equation in the case of sound waves, and so on — which describe the phenomena in which we are interested in a satisfactory manner. We should, however, like to write these equations of motion in canonical form, that is, we should like to find a new set of variables and a Hamiltonian which is a function of these variables such that the equations of motion can be written in the canonical form (5.108). Having found both the variables and the Hamiltonian, we can then tackle the problem as to how to introduce a canonical formalism when using the original variables which are functions of a continuous variable.

One-dimensional longitudinal elastic waves are described by the wave equation

$$\rho\ddot{\xi} - E\frac{\partial^2 \xi}{\partial x^2} = 0, \tag{8.101}$$

where $\xi(x, t)$ is the displacement at time t at the point x in the medium, ρ is the density, and E the Young modulus of the medium. We assume that the medium has a finite length L. We could require ξ (or $\partial\xi/\partial x$) to vanish at the boundaries, and we could then expand it in a Fourier series

$$\xi(x, t) = \sum_k \xi'_k(t) \sin kx, \tag{8.102}$$

where k can take on the values $n\pi/L$ ($n = 0, 1, 2, \ldots$). It is, however, more convenient to impose periodic boundary conditions upon ξ,

$$\xi(x+L, t) = \xi(x, t), \tag{8.103}$$

rather than the condition that ξ (or $\partial\xi/\partial x$) should vanish at the ends of the rod. One should emphasise here that there is little justification for (8.103); it is used because it is more convenient and usually the actual boundary conditions will not influence the problem so that we can choose them to suit our convenience. Also, it is more convenient to use instead of the expansion (8.102) one in terms of exponentials

$$\xi(x, t) = L^{-\frac{1}{2}} \sum_k \xi_k(t) e^{ikx}, \tag{8.104}$$

where the wavenumber k can be both positive and negative. The drawback of expansion (8.104) is that the ξ_k are now complex quantities while the ξ'_k of equation (8.102) could be chosen to be real. This has as a consequence that in the Hamiltonian (8.112) and the Lagrangian (8.110) instead of the squares of ξ_k, $\dot{\xi}_k$, ... products such as $\xi_k \xi_{-k}$ occur.

We note here that as $\xi(x, t)$ is a real quantity, the ξ_k and ξ_{-k} are not two independent complex variables (that is, corresponding to four independent variables), but satisfy the relation

$$\xi_k = \xi^*_{-k}, \tag{8.105}$$

so that there are as many independent variables as there are k-values.

The ξ_k can be obtained from $\xi(x)$ in the usual way

$$\xi_k = L^{-\frac{1}{2}} \int \xi(x) e^{-ikx} dx. \tag{8.106}$$

If we wish to go over to a continuum of k-values we let L tend to infinity and take the following limits

$$\sum_k \to \frac{L}{2\pi} \int dk, \qquad \xi_k \to \left(\frac{2\pi}{L}\right)^{\frac{1}{2}} \xi(k), \tag{8.107}$$

which leads to the well-known equations of Fourier analysis

$$\begin{aligned} \xi(x, t) &= (2\pi)^{-\frac{1}{2}} \int \xi(k, t) e^{ikx} dk, \\ \xi(k, t) &= (2\pi)^{-\frac{1}{2}} \int \xi(x, t) e^{-ikx} dx. \end{aligned} \tag{8.108}$$

Taking the Fourier transform of equation (8.101) we find

$$\rho\ddot{\xi}_k + k^2 E \xi_k = 0, \tag{8.109}$$

which are the equations of motion of our system — a system with an infinite number of degrees of freedom.

These equations can be derived from the following Lagrangian

$$L(\xi_k, \dot{\xi}_k) = \tfrac{1}{2}\rho \sum_k \dot{\xi}_k \dot{\xi}_{-k} - \tfrac{1}{2}E \sum_k k^2 \xi_k \xi_{-k}. \tag{8.110}$$

From the Lagrangian equations of motion (2.308) with the Lagrangian (8.110) and the ξ_k as the generalised coordinates we find, indeed, the equations of motion (8.109). It is of interest to note that the equation of motion involving ξ_k derives from the Lagrangian equation involving ξ_{-k}.

From the Lagrangian (8.110) we can in the usual way find the Hamiltonian. First of all we introduce the momentum π_k conjugate to ξ_k by the relation [see (2.310)]

$$\pi_k = \frac{\partial L}{\partial \dot{\xi}_k} = \rho \dot{\xi}_{-k}, \tag{8.111}$$

where we have to take into account that because the sum over k involves both negative and positive values of the components of k the term $\dot{\xi}_k \xi_{-k}$ occurs twice. From (5.104′) we now find the Hamiltonian

$$H(\xi_k, \pi_k) = \sum_k \pi_k \dot{\xi}_k - L = \frac{1}{2\rho} \sum_k \pi_k \pi_{-k} + \tfrac{1}{2}E \sum_k k^2 \xi_k \xi_{-k}, \tag{8.112}$$

and the canonical equations of motion (5.108) lead again to (8.109).

We know that the equations of motion can be derived from Hamilton's variational principle. This means for the case under consideration that (8.109) can be derived from the variational principle

$$\delta \int L dt = 0 \tag{8.113}$$

with L given by (8.110).

We shall now investigate the problem of how to modify the various equations when we make the transition from the ξ_k to $\xi(x)$. Our procedure must be such that the equations of motion reduce to (8.101). The transition can be performed by means of the relations (8.107) and (8.108) after the length of the one-dimensional system, L, has been allowed to tend to infinity.

We shall consider first of all the Hamiltonian and the Lagrangian, and discuss separately the two sums occurring in each. Using the relation (8.111) we see that the first sum in each case is of the form

$$\begin{aligned}\tfrac{1}{2}\rho \sum_k \dot{\xi}_k \dot{\xi}_{-k} &= \tfrac{1}{2}\rho \int \dot{\xi}(k)\dot{\xi}(-k)\mathrm{d}k \\ &= \tfrac{1}{2}\rho(2\pi)^{-\frac{1}{2}} \int \mathrm{d}x \dot{\xi}(x) \int \mathrm{d}k\, \mathrm{e}^{-\mathrm{i}kx}\dot{\xi}(-k) \\ &= \tfrac{1}{2}\rho \int \mathrm{d}x \dot{\xi}(x)\dot{\xi}(x) = \int \mathscr{T}(x)\mathrm{d}x, \end{aligned} \tag{8.114}$$

where

$$\mathscr{T}(x) = \tfrac{1}{2}\rho \dot{\xi}(x)\dot{\xi}(x) \tag{8.115}$$

is the *kinetic energy density*, that is, the kinetic energy per unit volume.

We turn now to the second sum and find

$$\begin{aligned}\tfrac{1}{2}E \sum_k k^2 \xi_k \xi_{-k} &= \tfrac{1}{2}E \int k^2 \xi(k)\xi(-k)\mathrm{d}k \\ &= \tfrac{1}{2}E(2\pi)^{-\frac{1}{2}} \iint \mathrm{d}x\,\mathrm{d}k\, \xi(x) k^2 \mathrm{e}^{-\mathrm{i}kx}\xi(-k) \\ &= \tfrac{1}{2}E(2\pi)^{-\frac{1}{2}} \iint \mathrm{d}x\,\mathrm{d}k\, \xi(x) \left(-\frac{\partial^2}{\partial x^2}\mathrm{e}^{-\mathrm{i}kx}\right)\xi(-k) \\ &= \tfrac{1}{2}E(2\pi)^{-\frac{1}{2}} \int \mathrm{d}x \left[-\frac{\partial^2}{\partial x^2}\xi(x)\right] \int \mathrm{d}k\,\mathrm{e}^{-\mathrm{i}kx}\xi(-k) \\ &= \tfrac{1}{2}E \int \mathrm{d}x \left[-\frac{\partial^2}{\partial x^2}\xi(x)\right]\xi(x) \\ &= \tfrac{1}{2}E \int \left(\frac{\partial \xi}{\partial x}\right)^2 \mathrm{d}x = \int \mathscr{U}(x)\mathrm{d}x, \end{aligned} \tag{8.116}$$

where

$$\mathscr{U}(x) = \tfrac{1}{2}E \left(\frac{\partial \xi}{\partial x}\right)^2 \tag{8.117}$$

is the potential energy density. This follows from the usual definition of the elastic (potential) energy density which is one half times the elastic modulus (here E) multiplied by the square of the appropriate strain component (in our case this component is the elongation per unit length, that is, $\partial\xi/\partial x$).

The transformation from the k- to the x-representation in the equations of motion (2.308) or (5.108) is not so easily made. The easiest way is, however, to remember that these equations follow in a straightforward manner from (8.113). As this is true when we use the ξ_k as our variables, it should remain to be true when we use $\xi(x)$ instead of the ξ_k. In the Lagrangian L as expressed in terms of $\xi(x, t)$ we shall now find not only $\xi(x, t)$ and $\dot{\xi}(x, t)$, but also $\partial\xi/\partial x$ [see (8.117)], and the variation of L will involve the variation of ξ, $\dot{\xi}$, and $\partial\xi/\partial x$, where the last two variations will not be independent of the first one [compare the derivation of (2.308)]. Introducing the *Lagrangian density* $\mathscr{L}$ by the equation

$$\mathscr{L} = \mathscr{T} - \mathscr{U}, \tag{8.118}$$

equation (8.113) can be written as

$$\delta \iint \mathscr{L} \mathrm{d}x \mathrm{d}t = 0. \tag{8.119}$$

We note that in (8.119) the space and time coordinates enter on the same basis. We should therefore expect that the variational principle in the form of (8.119) would be especially suitable for relativistic considerations and this is, indeed, the case.

The Lagrangian density $\mathscr{L}$ will be a function of ξ, $\dot{\xi}$, and $\partial\xi/\partial x$, and from (8.119) we get, taking into account that $\delta\dot{\xi}$ and $\delta\partial\xi/\partial x$ are not independent variations,

$$\delta \iint \mathscr{L} \mathrm{d}x \mathrm{d}t = \iint \delta\mathscr{L} \mathrm{d}x \mathrm{d}t$$

$$= \iint \left(\frac{\partial\mathscr{L}}{\partial\xi} \partial\xi + \frac{\partial\mathscr{L}}{\partial\dot{\xi}} \delta\dot{\xi} + \frac{\partial\mathscr{L}}{\partial\frac{\partial\xi}{\partial x}} \delta \frac{\partial\xi}{\partial x} \right) \mathrm{d}x \mathrm{d}t$$

$$= \iint \left[\frac{\partial\mathscr{L}}{\partial\xi} - \frac{\partial}{\partial t}\frac{\partial\mathscr{L}}{\partial\dot{\xi}} - \frac{\partial}{\partial x}\frac{\partial\mathscr{L}}{\partial\frac{\partial\xi}{\partial x}} \right] \delta\xi \mathrm{d}x \mathrm{d}t$$

$$+ \int \frac{\partial\mathscr{L}}{\partial\dot{\xi}} \delta\xi \mathrm{d}x + \int \frac{\partial\mathscr{L}}{\partial\frac{\partial\xi}{\partial x}} \delta\xi \mathrm{d}t$$

or

$$0 = \iint \left[\frac{\partial \mathscr{L}}{\partial \xi} - \frac{\partial}{\partial t} \frac{\partial \mathscr{L}}{\partial \dot{\xi}} - \frac{\partial}{\partial x} \frac{\partial \mathscr{L}}{\partial \dfrac{\partial \xi}{\partial x}} \right] \delta\xi \, \mathrm{d}x \, \mathrm{d}t, \tag{8.120}$$

where we have used the fact that $\delta\xi$ vanishes at the end points of both the time interval and of the space interval. We get thus the equations of motion

$$\frac{\partial}{\partial t} \frac{\partial \mathscr{L}}{\partial \dot{\xi}} - \frac{\partial \mathscr{L}}{\partial \xi} + \frac{\partial}{\partial x} \frac{\partial \mathscr{L}}{\partial \dfrac{\partial \xi}{\partial x}} = 0. \tag{8.121}$$

One sees easily that for $\mathscr{L}$ given by (8.115), (8.117), and (8.118),

$$\mathscr{L} = \tfrac{1}{2}\rho\dot{\xi}^2 - \tfrac{1}{2}E \left(\frac{\partial \xi}{\partial x}\right)^2, \tag{8.122}$$

equation (8.121) reduces to (8.101).

If we introduce the *functional derivatives* $\delta/\delta\xi$ of a function by the equation

$$\frac{\delta f\left(\xi, \dfrac{\partial \xi}{\partial x}\right)}{\delta \xi} \equiv \frac{\partial f}{\partial \xi} - \frac{\partial}{\partial x} \frac{\partial f}{\partial \dfrac{\partial \xi}{\partial x}}, \tag{8.123}$$

we can write (8.121) in the form

$$\frac{\partial}{\partial t} \frac{\partial \mathscr{L}}{\partial \dot{\xi}} - \frac{\delta \mathscr{L}}{\delta \xi} = 0, \tag{8.124}$$

which is formally very similar to the Lagrangian equations of motion (2.308). We must emphasise, however, that the Lagrangian density $\mathscr{L}$ enters into (8.124) while (2.308) contains the total Lagrangian L.

We can now introduce a canonical momentum density π by the equation

$$\pi = \frac{\partial \mathscr{L}}{\partial \dot{\xi}}, \tag{8.125}$$

and a *Hamiltonian density* $\mathscr{H}$ by the relation [compare (5.104′)]

$$\mathscr{H} = \pi\dot{\xi} - \mathscr{L}. \tag{8.126}$$

From the variational principle (8.119) we now find

$$\iint \delta(\pi\dot{\xi} - \mathscr{H}) \mathrm{d}x \, \mathrm{d}t = 0, \tag{8.127}$$

and taking into account that $\mathscr{H}$ is a function of π, ξ, and $\partial\xi/\partial x$ we get by a method similar to the one used to derive equation (8.121)

$$\iint \left[\left(\dot{\xi} - \frac{\partial \mathscr{H}}{\partial \pi} \right) \delta\pi - \left(\dot{\pi} + \frac{\delta \mathscr{H}}{\delta \xi} \right) \delta\xi \right] \mathrm{d}x\mathrm{d}t = 0, \tag{8.128}$$

whence follows

$$\dot{\xi} = \frac{\delta \mathscr{H}}{\delta \pi}, \qquad \dot{\pi} = -\frac{\delta \mathscr{H}}{\delta \xi}, \tag{8.129}$$

where we have replaced $\partial\mathscr{H}/\partial\pi$ by the functional derivative $\delta\mathscr{H}/\delta\pi$ to get symmetric equations. We note that equations (8.129) are different from the previous canonical equations (5.108) in three respects: functional derivatives appear instead of ordinary partial derivatives, the Hamiltonian density instead of the total Hamiltonian, and the momentum density instead of the momentum.

If we apply this canonical formalism to the case of the one-dimensional elastic waves, we find from (8.125)

$$\pi = \rho\dot{\xi}, \tag{8.130}$$

and comparing this with (8.111) and using (8.106) we see that the π_k are the Fourier components of the momentum density π corresponding to the wave number $-k$. Equation (8.126) now leads to the Hamiltonian density

$$\mathscr{H} = \frac{\pi^2}{2\rho} + \tfrac{1}{2}E\left(\frac{\partial\xi}{\partial x}\right)^2 = \mathscr{T} + \mathscr{U}, \tag{8.131}$$

which could also have been derived from (8.112) using equations (8.130), (8.115), and (8.117). We notice that in this case the Hamiltonian density is the same as the total energy density.

From the first of equations (8.129) we get

$$\dot{\xi} = \pi/\rho, \tag{8.132}$$

which is the same as (8.130) while the second of equations (8.129) leads to

$$\dot{\pi} = E\frac{\partial^2\xi}{\partial x^2}. \tag{8.133}$$

Equation (8.101) follows now immediately if we combine (8.132) and (8.133).

The extension of the theory developed here to the case of several variables and of three-dimensional systems is straightforward. For instance, in the case of sound waves in a three-dimensional medium the components ξ, η, and ζ of the displacement vector $\boldsymbol{\xi}(\boldsymbol{x})$ will be functions of x, y, and z, and the Lagrangian will be a function of ξ, η, ζ, $\dot{\xi}$, $\dot{\eta}$, $\dot{\zeta}$, $\partial\xi/\partial x$, $\partial\xi/\partial y$, $\partial\xi/\partial z$, $\partial\eta/\partial x$, $\partial\eta/\partial y$, $\partial\eta/\partial z$, $\partial\zeta/\partial x$, $\partial\zeta/\partial y$, and $\partial\zeta/\partial z$. For each of the three components we shall have a Lagrangian equation of the form (8.124) while the functional derivatives are now defined by equations of the form

$$\frac{\delta f}{\delta \xi} \equiv \frac{\partial f}{\partial \xi} - \frac{\partial}{\partial x}\frac{\partial f}{\partial \dfrac{\partial \xi}{\partial x}} - \frac{\partial}{\partial y}\frac{\partial f}{\partial \dfrac{\partial \xi}{\partial y}} - \frac{\partial}{\partial z}\frac{\partial f}{\partial \dfrac{\partial \xi}{\partial z}}. \tag{8.134}$$

In the next section we shall consider applications both of the formalism involving functional derivatives and of the formalism involving Fourier components.

2. SOUND WAVES; THE MAXWELL EQUATIONS

A physical system may be defined by the equations of motion it must satisfy, or it may be defined by its Lagrangian. The former case has occurred in the systems discussed in this book, but the latter case is often encountered in field theory. In this section we shall shew that the choice of certain Lagrangian densities will lead to the equations of motion for the system under consideration, but we shall not derive the Lagrangians from the equations of motion as was done in earlier chapters and also in the preceding section. We shall consider first sound waves and then the Maxwell equations starting from an appropriate Lagrangian density. After that we shall discuss the Maxwell equations using the method of Fourier components.

We shall prove that the sound wave equation

$$\ddot{\rho} - s^2\nabla^2\rho = 0 \tag{8.201}$$

can be derived from the Lagrangian density

$$\mathscr{L} = \tfrac{1}{2}(\dot{\boldsymbol{\xi}} \cdot \dot{\boldsymbol{\xi}}) - \tfrac{1}{2}s^2(\nabla \cdot \boldsymbol{\xi})^2. \tag{8.202}$$

In these equations ρ is the density in the system, the (constant) equilibrium value of which is ρ_0, s the sound velocity, and $\boldsymbol{\xi}$ the displacement vector with components ξ, η, and ζ. The density ρ and $\boldsymbol{\xi}$ are related through the equation of continuity

$$\dot{\rho}+\rho_0(\nabla\cdot\dot{\xi}) = 0, \quad \text{or,} \quad \frac{\rho-\rho_0}{\rho_0} = -(\nabla\cdot\xi). \tag{8.203}$$

From (8.202) we find

$$\frac{\partial\mathscr{L}}{\partial\dot{\xi}} = \dot{\xi}, \qquad \frac{\partial\mathscr{L}}{\partial\dot{\eta}} = \dot{\eta}, \qquad \frac{\partial\mathscr{L}}{\partial\dot{\zeta}} = \dot{\zeta};$$

$$\frac{\partial\mathscr{L}}{\partial\xi} = \frac{\partial\mathscr{L}}{\partial\eta} = \frac{\partial\mathscr{L}}{\partial\zeta} = 0;$$

$$\frac{\partial\mathscr{L}}{\partial\dfrac{\partial\xi}{\partial x}} = -s^2(\nabla\cdot\xi) = \frac{\partial\mathscr{L}}{\partial\dfrac{\partial\eta}{\partial y}} = \frac{\partial\mathscr{L}}{\partial\dfrac{\partial\zeta}{\partial z}}, \tag{8.204}$$

$$\frac{\partial\mathscr{L}}{\partial\dfrac{\partial\xi}{\partial y}} = \frac{\partial\mathscr{L}}{\partial\dfrac{\partial\xi}{\partial z}} = \frac{\partial\mathscr{L}}{\partial\dfrac{\partial\eta}{\partial x}} = \frac{\partial\mathscr{L}}{\partial\dfrac{\partial\eta}{\partial z}} = \frac{\partial\mathscr{L}}{\partial\dfrac{\partial\zeta}{\partial x}} = \frac{\partial\mathscr{L}}{\partial\dfrac{\partial\zeta}{\partial y}} = 0.$$

These equations lead to

$$\frac{\delta\mathscr{L}}{\delta\xi} = s^2(\nabla\cdot\xi) = \frac{\delta\mathscr{L}}{\delta\eta} = \frac{\delta\mathscr{L}}{\delta\zeta}, \tag{8.205}$$

and we get from (8.124)

$$\ddot{\xi}-s^2\nabla(\nabla\cdot\xi) = 0, \tag{8.206}$$

which reduces to (8.201) if we take the divergence of this equation and use (8.203).

From (8.204) and (8.125) we find for the momentum density vector

$$\boldsymbol{\pi} = \dot{\xi}, \tag{8.207}$$

and from (8.126), for the Hamiltonian

$$\mathscr{H} = \tfrac{1}{2}\boldsymbol{\pi}\cdot\boldsymbol{\pi}+\tfrac{1}{2}s^2(\nabla\cdot\xi)^2. \tag{8.208}$$

Equations (8.129) lead again to (8.206).

To conclude this section we shall consider the electromagnetic field. The magnetic induction $\boldsymbol{B}$, the magnetic field strength $\boldsymbol{H}$, the electrical displacement vector $\boldsymbol{D}$, and the electrical field strength $\boldsymbol{E}$ satisfy in vacuo the following equations

$$\boldsymbol{B} = \mu_0\boldsymbol{H}, \qquad \boldsymbol{D} = \varepsilon_0\boldsymbol{E}, \tag{8.209}$$

and the Maxwell equations

$$\text{(i)}\quad (\nabla \cdot \boldsymbol{D}) = \rho; \qquad \text{(ii)}\quad (\nabla \cdot \boldsymbol{B}) = 0;$$
$$\text{(iii)}\quad [\nabla \wedge \boldsymbol{E}] = -\frac{\partial \boldsymbol{B}}{\partial t}; \qquad \text{(iv)}\quad [\nabla \wedge \boldsymbol{H}] = \boldsymbol{j} + \frac{\partial \boldsymbol{D}}{\partial t}, \tag{8.210}$$

where ρ and $\boldsymbol{j}$ are here the electrical charge and current densities, and where μ_0 and ε_0 are the permeability and permittivity of free space, respectively. We introduce again the vector potential $\boldsymbol{A}$ and scalar potential ϕ by the equations [see (2.506)]

$$\boldsymbol{B} = [\nabla \wedge \boldsymbol{A}], \qquad \boldsymbol{E} = -\nabla\phi - \frac{\partial \boldsymbol{A}}{\partial t}, \tag{8.211}$$

and we treat ϕ and $\boldsymbol{A}$ as the $Q(\boldsymbol{x})$ of this system. The Lagrangian density of the electromagnetic field is given by the equation (c: velocity of light in vacuo)

$$\mathscr{L} = \tfrac{1}{2}\varepsilon_0(\boldsymbol{E}^2 - c^2\boldsymbol{B}^2) - \rho\phi + (\boldsymbol{j} \cdot \boldsymbol{A}). \tag{8.212}$$

From this expression we find

$$\frac{\partial \mathscr{L}}{\partial \phi} = -\rho, \qquad \frac{\partial \mathscr{L}}{\partial \dot{\phi}} = 0,$$
$$\frac{\partial \mathscr{L}}{\partial \dfrac{\partial \phi}{\partial x}} = -\varepsilon_0 E_x, \qquad \frac{\partial \mathscr{L}}{\partial \dfrac{\partial \phi}{\partial y}} = -\varepsilon_0 E_y, \qquad \frac{\partial \mathscr{L}}{\partial \dfrac{\partial \phi}{\partial z}} = -\varepsilon_0 E_z, \tag{8.213}$$

and (8.210 i) follows as a Lagrangian equation of motion.

We also have

$$\frac{\partial \mathscr{L}}{\partial A_x} = j_x, \qquad \frac{\partial \mathscr{L}}{\partial \dot{A}_x} = \varepsilon_0 \left(\boldsymbol{E} \cdot \frac{\partial \boldsymbol{E}}{\partial \dot{A}_x}\right) = -D_x,$$
$$\frac{\partial \mathscr{L}}{\partial \dfrac{\partial A_x}{\partial x}} = 0, \qquad \frac{\partial \mathscr{L}}{\partial \dfrac{\partial A_x}{\partial y}} = H_z, \qquad \frac{\partial \mathscr{L}}{\partial \dfrac{\partial A_x}{\partial z}} = -H_y, \tag{8.214}$$

where we have used equations (8.211) and the relation $\varepsilon_0 \mu_0 = c^{-2}$. From (8.214) and (8.124), equation (8.210 iv) follows. Equations (8.210 ii and iii) are satisfied, of course, by the introduction of the potentials $\boldsymbol{A}$ and ϕ through equations (8.211).

As $\mathscr{L}$ does not contain $\dot{\phi}$ we cannot introduce a momentum density corresponding to ϕ, and it is impossible without further alterations to find a Hamiltonian density such that equations (8.210) follow from equations (8.129). We shall, however, see that one can write the Maxwell equations in canonical form by introducing the Fourier components of the field variables.

We first of all introduce a complex variable $\boldsymbol{F}$ by the equation

$$\boldsymbol{F} = \boldsymbol{E} + \mathrm{i}c\boldsymbol{B}. \tag{8.215}$$

We can thus write equations (8.210) in the form

$$(\nabla \cdot \boldsymbol{F}) = \frac{\rho}{\varepsilon_0} \quad \text{(i)}; \qquad [\nabla \wedge \boldsymbol{F}] - \frac{\mathrm{i}}{c}\frac{\partial \boldsymbol{F}}{\partial t} = \mathrm{i}\sqrt{\frac{\mu_0}{\varepsilon_0}}\,\boldsymbol{j}. \quad \text{(ii)} \tag{8.216}$$

We now introduce Fourier expansions, and at the same time decompose each term into three components, one parallel to the wave vector $\boldsymbol{k}$ and the other two at right angles to $\boldsymbol{k}$:

$$\boldsymbol{F} = \Omega^{-\frac{1}{2}} \sum_{\boldsymbol{k}} \mathrm{e}^{\mathrm{i}(\boldsymbol{k}\cdot\boldsymbol{x})}(a_{\boldsymbol{k}}\boldsymbol{e}_{\boldsymbol{k}}^{(1)} + b_{\boldsymbol{k}}\boldsymbol{e}_{\boldsymbol{k}}^{(2)} + c_{\boldsymbol{k}}\boldsymbol{e}_{\boldsymbol{k}}^{(3)}), \tag{8.217}$$

where the $\boldsymbol{e}_{\boldsymbol{k}}^{(i)}$ are unit vectors which form a right-handed coordinate system (we assume that the x, y, z-system is right-handed) and where $\boldsymbol{e}_{\boldsymbol{k}}^{(3)}$ is parallel to $\boldsymbol{k}$, so that

$$\begin{aligned} &[\boldsymbol{e}_{\boldsymbol{k}}^{(1)} \wedge \boldsymbol{e}_{\boldsymbol{k}}^{(2)}] = \boldsymbol{e}_{\boldsymbol{k}}^{(3)} \text{ (cyclic)}, \quad [\boldsymbol{k} \wedge \boldsymbol{e}_{\boldsymbol{k}}^{(1)}] = k\boldsymbol{e}_{\boldsymbol{k}}^{(2)}, \quad [\boldsymbol{k} \wedge \boldsymbol{e}_{\boldsymbol{k}}^{(2)}] = -k\boldsymbol{e}_{\boldsymbol{k}}^{(1)}, \\ &[\boldsymbol{k} \wedge \boldsymbol{e}_{\boldsymbol{k}}^{(3)}] = 0, \qquad (\boldsymbol{k}\cdot\boldsymbol{e}_{\boldsymbol{k}}^{(1)}) = (\boldsymbol{k}\cdot\boldsymbol{e}_{\boldsymbol{k}}^{(2)}) = 0, \qquad (\boldsymbol{k}\cdot\boldsymbol{e}_{\boldsymbol{k}}^{(3)}) = k, \end{aligned} \tag{8.218}$$

where $k = |\boldsymbol{k}|$; finally Ω is the finite volume at the boundaries of which periodic boundary conditions are assumed to hold. One should note that the decomposition using $\boldsymbol{e}_{\boldsymbol{k}}^{(1)}$, $\boldsymbol{e}_{\boldsymbol{k}}^{(2)}$, and $\boldsymbol{e}_{\boldsymbol{k}}^{(3)}$ is different for different $\boldsymbol{k}$, that is for different terms in the sum over wave vectors.

Assuming similar expansions for ρ and $\boldsymbol{j}$

$$\rho = \Omega^{-\frac{1}{2}} \sum_{\boldsymbol{k}} \rho_{\boldsymbol{k}} \mathrm{e}^{\mathrm{i}(\boldsymbol{k}\cdot\boldsymbol{x})}, \tag{8.219}$$

$$\boldsymbol{j} = \Omega^{-\frac{1}{2}} \sum_{\boldsymbol{k}} \mathrm{e}^{\mathrm{i}(\boldsymbol{k}\cdot\boldsymbol{x})}(j_{\boldsymbol{k}}^{(1)}\boldsymbol{e}_{\boldsymbol{k}}^{(1)} + j_{\boldsymbol{k}}^{(2)}\boldsymbol{e}_{\boldsymbol{k}}^{(2)} + j_{\boldsymbol{k}}^{(3)}\boldsymbol{e}_{\boldsymbol{k}}^{(3)}), \tag{8.220}$$

we get from equations (8.216) the following relations

$$\mathrm{i}kc_{\boldsymbol{k}} = \rho_{\boldsymbol{k}}/\varepsilon_0, \tag{8.221}$$

$$\begin{aligned} -ikb_k - i\dot{a}_k c^{-1} &= iR_0 j_k^{(1)}, \quad \text{(i)} \\ ika_k - i\dot{b}_k c^{-1} &= iR_0 j_k^{(2)}, \quad \text{(ii)} \\ -i\dot{c}_k c^{-1} &= iR_0 j_k^{(3)}, \quad \text{(iii)} \end{aligned} \tag{8.222}$$

where

$$R_0 = (\mu_0/\varepsilon_0)^{\frac{1}{2}} \tag{8.223}$$

is the characteristic impedance of free space.

From the equation of continuity,

$$(\nabla \cdot j) + \frac{\partial \rho}{\partial t} = 0, \tag{8.224}$$

we find

$$ikj_k^{(3)} = -\dot{\rho}_k, \tag{8.225}$$

and we see that (8.222 iii) follows directly from (8.221).

In the following we shall assume that there are no charges or currents so that $\rho_k = j_k^{(1)} = j_k^{(2)} = j_k^{(3)} = 0$, and therefore also from (8.221)

$$c_k = 0. \tag{8.226}$$

The electromagnetic field is now a transverse field (a radiation field) and is described by the (complex) variables a_k and b_k. As they are complex we can take either their real and imaginary parts as independent variables, or a_k, a_k^*, b_k, b_k^* as independent variables. The situation is different here from the one at the beginning of the preceding section which led to (8.105), as F is a complex quantity while ξ was real. The equations of motion governing these variables are [compare (8.222 i and ii)]

$$\begin{aligned} \dot{a}_k &= -ckb_k, & \dot{a}_k^* &= -ckb_k^*, \\ \dot{b}_k &= cka_k, & \dot{b}_k^* &= cka_k^*. \end{aligned} \tag{8.227}$$

We shall use here as independent variables the real and imaginary parts of a_k and b_k, and write

$$a_k = \varepsilon_0^{-\frac{1}{2}}(p_k + i\bar{p}_k), \qquad b_k = kc\varepsilon_0^{-\frac{1}{2}}(q_k + i\bar{q}_k), \tag{8.228}$$

so that equations (8.227) are of the form

$$\dot{p}_k = -(ck)^2 q_k, \quad \dot{q}_k = p_k, \quad \dot{\bar{p}}_k = -(ck)^2 \bar{q}_k, \quad \dot{\bar{q}}_k = \bar{p}_k. \tag{8.229}$$

These equations can be derived from the canonical equations of motion (5.108) with a Hamiltonian

$$H = \tfrac{1}{2}\sum_{\boldsymbol{k}} [p_{\boldsymbol{k}}^2 + (ck)^2 q_{\boldsymbol{k}}^2 + \bar{p}_{\boldsymbol{k}}^2 + (ck)^2 \bar{q}_{\boldsymbol{k}}^2], \tag{8.230}$$

if $p_{\boldsymbol{k}}$, $q_{\boldsymbol{k}}$ and $\bar{p}_{\boldsymbol{k}}$, $\bar{q}_{\boldsymbol{k}}$ are treated as pairs of canonically conjugate variables.

If we express H in terms of the $a_{\boldsymbol{k}}$, $a_{\boldsymbol{k}}^*$, $b_{\boldsymbol{k}}$, and $b_{\boldsymbol{k}}^*$, we get

$$H = \tfrac{1}{2}\varepsilon_0 \sum_{\boldsymbol{k}} (a_{\boldsymbol{k}} a_{\boldsymbol{k}}^* + b_{\boldsymbol{k}} b_{\boldsymbol{k}}^*), \tag{8.231}$$

and if we take as canonical momenta $\gamma_{\boldsymbol{k}} a_{\boldsymbol{k}}$ and $\gamma_{\boldsymbol{k}} a_{\boldsymbol{k}}^*$, and as their conjugate coordinates $\gamma_{\boldsymbol{k}} b_{\boldsymbol{k}}^*$ and $\gamma_{\boldsymbol{k}} b_{\boldsymbol{k}}$ with $\gamma_{\boldsymbol{k}}^2 = \varepsilon_0/2ck$, equations (8.227) follow from equations (5.108).

We see from (8.230) that the problem of the electromagnetic radiation field can be reduced to that of a set of harmonic oscillators. We also note that expression (8.230) for H can be rewritten by means of (8.228), (8.226), (8.217), and (8.215) in the form

$$H = \tfrac{1}{2}\varepsilon_0 \int (\boldsymbol{E}^2 + c^2 \boldsymbol{B}^2) \mathrm{d}^3 \boldsymbol{x}, \tag{8.232}$$

the well-known expression for the energy of the electromagnetic field.

The case where $\boldsymbol{j}$ and ρ do not vanish can be treated in a similar fashion, but we leave this to the reader. There are some complications which are discussed in the literature [for instance, H. A. Kramers, Quantum Mechanics (North Holland Publishing Company, 1957) Ch. 8].

BIBLIOGRAPHY

We give here a short list of references for further reading. References 3, 7, 9, 10, and 13 are modern textbooks on classical mechanics at roughly the same level as the present textbook, and reference 7 should be consulted for a more extensive bibliography, which is partly annotated. References 15, and 17 are textbooks which, although published some time ago, still retain their usefulness. Readers who wish to test their knowledge by solving problems can find those — of a greatly varying degree of difficulty — in references 3, 7, 9, 13, and 17. References 4, 5, and 8 are concerned with the basic ideas and the historical development of classical mechanics, and references 11 and 16 with the use of classical mechanics in celestial mechanics. References 2 and 6 are given here as they deal with adiabatic invariants and references to this subject are rather difficult to unearth. References 1 and 12 are only interested in classical mechanics inasfar as it can explain the properties of atomic systems.

1. M. Born, The Mechanics of the Atom (Bell, London, 1927).
2. J. M. Burgers, Het Atoommodel van Rutherford-Bohr (Loosjes, Haarlem, 1918); Versl. Kon. Akad. Wet., Amsterdam, **25** (1916/7) 849, 918, 1055; Ann. Phys. (Lpz.) **52** (1917) 195.
3. H. C. Corben and P. Stehle, Classical Mechanics (John Wiley, New York, 1950).
4. J. L. Destouches, Principes de la Mécanique Classique (C.N.R.S., Paris, 1948).
5. R. Dugas, Histoire de la Mécanique (Griffon, Neuchâtel, 1950).
6. P. Ehrenfest, Collected Scientific Papers (North Holland, Amsterdam, 1959) Paper no. 37.
7. H. Goldstein, Classical Mechanics (Addison-Wesley, Cambridge, Mass., 1950).
8. Handbuch der Physik, Volume V, Grundlagen der Mechanik, Mechanik der Punkte und Starrer Körper. (Springer, Berlin, 1927).

9. L. D. Landau and E. M. Lifshitz, Mechanics (Pergamon Press, London 1960).
10. J. W. Leech, Classical Mechanics (Methuen, London, 1958).
11. H. Poincaré, Les Méthodes Nouvelles de la Mécanique Céleste (Gauthier-Villars, Paris, 1892–99).
12. A. Sommerfeld, Atomic Structure and Spectral Lines (Dutton, New York, 1934).
13. A. Sommerfeld, Mechanics (Academic Press, New York, 1950).
14. R. C. Tolman, The Principles of Statistical Mechanics (Oxford University Press, 1938).
15. A. G. Webster, The Dynamics of Particles and of Rigid, Elastic and Fluid Bodies (Teubner, Leipzig, 1904).
16. A. Wintner, The Analytical Foundations of Celestial Mechanics (Princeton University Press, 1941).
17. E. T. Whittaker, Analytical Dynamics (Cambridge University Press, 1937).

PROBLEMS

CHAPTER 1

1. Use Binet's method to find the orbit corresponding to $U = -Ar^{-1} - Br^{-2}$. Discuss this orbit.

2. Discuss the motion of a yo-yo, that is, a rigid body which has the form of a disc and which moves up and down along a string which is wound on and unwound from a cylindrical rod, coaxial with the disc.

3. A rocket, initially of total mass M, throws off every second a mass αM with constant velocity V. Shew that it cannot rise at once unless $\alpha V > g$ and that it cannot rise at all unless $\alpha V > \beta g$ where βM is the mass of the case of the rocket.

Calculate the greatest height it can reach for the case where the conditions are such that the rocket is just able to rise vertically at once.

4. A spherical raindrop of constant density ρ falls vertically under gravity and a resistance proportional to its speed. Its radius increases uniformly due to condensation. Determine the speed of the drop and shew that the acceleration eventually attains a limiting value.

5. A particle of unit mass is moving in a potential energy field where $U = -\mu r^{-n}$.

(i) Shew that the particle can be projected with finite velocity from any point A, other than the origin, so as to reach infinity if and only if $n > 0$.

(ii) Evaluate the escape velocity as function of the distance a of A from the origin.

(iii) A particle is projected from A with escape velocity. Shew that it has escape velocity at every point in its path; that its orbit, if curved, is a circle, if $n = 4$; and that its orbit, if curved, will not reach infinity, if $n > 2$.

6. A particle of unit mass is projected from a point $\boldsymbol{r}_0$ with velocity $\boldsymbol{v}$ along a line whose perpendicular distance from the origin is p. An attractive

force $-N\mathbf{r}/r^3$ acts upon the particle. Shew that, if $N/r_0 \ll v^2$, the particle is deflected approximately through an angle $\pi - 2 \arctan (v^2 p/N)$.

7. Derive without using (1.255) the relation (1.258) between the scattering angle and the impact parameter, using (i) the conservation of momentum and (ii) the fact that the change in linear momentum along the original direction of the particle is obtained by integrating the appropriate component of the force.

8. Evaluate the differential cross-section for the scattering of a point particle by a hard-sphere potential.

9. Express the differential cross-section obtained in the preceding problem in terms of the energy lost by the scattered particle.

CHAPTER 2

1. Use the Lagrangian equations of the first kind to discuss the spherical pendulum.

2. Use the Lagrangian equations of the first kind to discuss the motion of a mass point on an inclined plane under the influence of the (homogeneous) gravitational field.

3. Use the Lagrangian equations of the first kind to discuss the motion of a particle which is constrained to move simultaneously on the surface of a sphere and in a given plane.

4. Discuss the motion of a hoop rolling down an inclined plane.

5. Use the Lagrangian equations of motion to discuss the motion of two particle connected by a flexible, inextensible, frictionless string. One particle moves on a smooth horizontal table, while the string passes through a small hole in that table.

6. A solid homogeneous cylinder of mass m and radius r rolls without slipping down an inclined plane face of a wedge of mass M, which in turn rests on a perfectly smooth horizontal floor. The angle between the inclined surface and the horizontal is ψ, and the entire motion takes place in a plane perpendicular to the floor and containing the normal to the inclined plane. Use the Lagrangian equations of motion to find the acceleration of the wedge.

7. A uniform ladder of weight Mg and length L has one end on a smooth

horizontal floor and the other end against a smooth vertical wall. The ladder is initially at rest in a vertical plane perpendicular to the wall and makes an angle of 60° with the horizontal. Integrate the Lagrangian equations of motion to find the motion of the ladder until it hits the floor.

8. A bead of mass m slides freely on a smooth circular wire of radius a which rotates about an axis through one of its points and perpendicular to its plane with angular velocity ω. Investigate the detailed motion of the bead and obtain an expression for the reaction of the wire against the bead.

CHAPTER 3

1. Discuss† the small vibrations in a vertical plane of a uniform straight rod which is suspended by a weightless inextensible string fixed at one of its ends.

2. Discuss the small vibrations in a vertical plane of a system consisting of a uniform straight rod which can freely rotate about a fixed point where one end is pivoted, while from its other end there hangs a mass point at the end of a weightless inextensible string.

3. A uniform circular disc is suspended from a fixed point by a weightless inextensible string which is attached to a point on its circumference. Discuss the small vibrations in a vertical plane of this system under gravity.

4. A smooth thin heavy wire bent in the form of a circle carries a bead. Discuss the small vibrations under gravity when the wire swings in its own plane about a point on the circle and the bead slides on the wire.

5. A weightless inextensible string of length $(2\sqrt{2}+1)a$ has its ends attached to points A and D a distance $3a$ apart and on the same horizontal level. Two particles, each of mass m, are attached to the string at B and C, respectively, where AB = CD = $\sqrt{2}\cdot a$. A particle of mass m is suspended by a weightless inextensible string of length a from B and another particle of mass m is similarly suspended from C. Discuss the small vibrations under gravity in the vertical plane containing the strings.

6. A weightless string of length $4a$ is stretched to tension T between two fixed points; three particles of mass μ are placed at the points of quadri-

† In this and subsequent problems "discuss" is meant to imply "find the eigenfrequencies of the normal vibrations and describe the normal modes".

section. Discuss, neglecting gravity, the small longitudinal and the small transverse vibrations.

7. A weightless string of length $4a$ is stretched to tension T between two fixed points A and B. A particle of mass m is attached to its midpoint C and two particles, each of mass m, are attached to it at the midpoints D and E of AC and CB. When the system is at rest, small transverse velocities u are suddenly given in the same directions to both particles at D and E. Find the displacement of the particle at C as a function of time.

8. A uniform rigid straight rod AB lies on a smooth horizontal table. The ends A and B are attached by taut weightless strings to fixed points C and D of the table. At rest CABD is a straight line. Discuss the small transverse vibrations of the rod.

9. Three particles of different masses are attached to points of a weightless inextensible string, one end of which is attached to a fixed point. Prove that if the periods of the three normal modes are the same as the periods of simple pendulums of lengths λ_1, λ_2, λ_3, then $\lambda_1+\lambda_2+\lambda_3$ is equal to the distance of the lowest particle from the point of suspension.

10. On each of two parallel horizontal wires a ring of mass m is kept in position by weightless springs, such that when the ring is displaced a distance y from its equilibrium position a restoring force λy acts upon it. When the two rings are in their equilibrium positions, the line joining them is at right angles to both wires and their distance apart is l. A uniform string, of line density ρ, is stretched to tension T between the two rings. Discuss the normal vibrations of this system.

11. A uniform rigid bar, of mass M, of length L, and of negligible width, is suspended symmetrically by two vertical springs, each of stiffness constant α, spaced a distance $x(< L)$ apart, so that at equilibrium the bar is horizontal. Show that, if $x = L/\sqrt{3}$, it is possible to set the bar into oscillation in the vertical plane so that an arbitrary point on the bar remains at rest.

12. A uniform string of length l and line density ρ lies on a smooth horizontal plane. One end is attached to a fixed point A on the plane, and the other to a ring of mass M which slides on a smooth horizontal rail in the plane at a distance l from A. The tension in the string is T. Find an equation for the period of small oscillations.

13. A metal block of mass $3m$ contains a spherical cavity of radius a in which slides, without rolling, a sphere of mass m whose radius is small compared to a. The block slides on a smooth horizontal plane and is attached to a fixed point by a horizontal spring whose force constant is $4mg/a$. Discuss the small vibrations of this system when it is moving along the direction of the spring.

14. A uniform thin hollow cylinder, of mass $2m$ and radius $2a$, rolls on a perfectly rough horizontal plane. Inside it rolls a second uniform thin hollow cylinder, of mass m and radius a, which is loaded at a point on its surface with a particle of mass $2m$. Discuss the small vibrations of this system, if there is no slipping between the cylinders.

15. A rhombus formed of four equal uniform rods of length $2L$, freely jointed, is placed symmetrically in a vertical plane over a smooth fixed cylinder of radius a and is in equilibrium when each rod is inclined at an angle ψ to the horizontal and the lower rods hang clear of the cylinder. Find the value of ψ and evaluate in terms of ψ the frequency of small vibrations of the rhombus when it is symmetrically disturbed.

16. A pocket watch rests on a smooth horizontal table. Assuming that it will perform oscillations with a small amplitude and that it keeps perfect time when the case of the watch is fixed, evaluate the rate of gain of the watch.

17. ABCD is a uniform square lamina of side $2a$. The corner A can slide smoothly along a fixed horizontal rod. The corner B is attached by a weightless inextensible string of length $\surd 3a$ to a fixed point such that, in equilibrium, the lamina hangs in a vertical plane perpendicular to the rod with the string vertical and CD horizontal and below AB. Discuss the small vibrations of the system.

18. Consider a one-dimensional 'crystal' of N atoms arranged on a circle (see § 3.4). If at $t = 0$, one of the atoms is displaced over a small distance d, describe the subsequent motion of the atoms in the crystal.

19. A thin wire of mass M is bent in the form of the helix $x = a \cos \psi$, $y = a \sin \psi$, $z = a\psi \tan \beta$, and is mounted so as to be free to rotate about the z-axis which is horizontal. The length of the wire is such that the centre of mass lies on the axis. A small smooth ring of mass m slides on the wire.

The wire is at rest when the ring is projected horizontally along the wire from the lowest position on it with velocity $a\omega$. Shew that if ω is not too large the plane through the axis and the ring oscillates like a simple pendulum of length

$$a(M+m\sin^2\beta)/(M\cos^2\beta+m\sin^2\beta),$$

and that the wire turns through an angle

$$m\omega T\sin^2\beta\cos\beta/(M\cos^2\beta+m\sin^2\beta),$$

in each period T of the pendulum motion.

20. A uniform rod AB, of length $2l$ and mass m, moves in a vertical plane under gravity, while one of its ends is constrained to move in a horizontal line LM with uniform acceleration a. Find the period of small oscillations of the rod about its relative equilibrium position.

21. Discuss the small vibrations around the equilibrium motion of the Thomson-Tait pendulum.

22. A particle is constrained to move on a smooth surface in the form of the anchor ring

$$x = \rho\cos\psi, \qquad y = \rho\sin\psi, \qquad z = b\sin\theta,$$

where $\rho = a+b\cos\theta$ $(a > b)$.

Prove that if there are no forces except the reaction of the surface, motion round the outer equatorial circle will be stable; and that if this motion is slightly disturbed, the new path will cross the equator at intervals of length $\pi\sqrt{b(a+b)}$.

Shew also that motion round the inner equatorial circle is unstable, and that if such motion is slightly disturbed the path of the particle will cross the outer equatorial circle at an angle $2\arctan\sqrt{b/a}$.

CHAPTER 4

1. A rigid body, free to rotate about a fixed point O is at rest when it is acted upon by an impulsive couple with components ε, ε, 1 along the principal axes of the body at O, ε being small. If $C > B > A$, shew that the inclination of the instantaneous axis of rotation to the C-axis in the ensuing motion is, to first order in ε, always less than or equal to $k\varepsilon$, where k is a number depending only on the ratios C/A and B/A. Find k explicitly when $C = 3A$ and $B = 2A$.

2. The principal moments of inertia of a rigid body about a fixed point are A, B, and C ($A < B < C$) and the corresponding components of the angular velocity are ω_1, ω_2, and ω_3. Shew that $A(C-A)\omega_1^2+B(C-B)\omega_2^2$ and $C(C-A)\omega_3^2+B(B-A)\omega_2^2$ are constants of motion. If these constants are in the ratio $C-B$ to $B-A$, and if $\omega_1 = \omega_0$, $\dot{\omega}_1 = 0$ at $t = 0$, find the time-dependence of ω_1.

3. Use a Routh function to ignore the cyclic coordinates for the case of a symmetric top rotating about a fixed point and hence derive the differential equation for the equation of motion for the non-cyclical coordinate.

4. If T is the kinetic energy of the rotational motion of a rigid body, and J the magnitude of the angular momentum, (Compare eqns. (4.208) and (4.209)), and if $J^2 = 2TC(1+\varepsilon)$, where ε is small, and $A > B > C$, shew that ω_1 and ω_2 undergo a simple harmonic variation. Calculate the period of this motion, and find correct to terms in $\sqrt{\varepsilon}$ the amplitude of ω_1.

5. Consider a top of mass M, height of centre of mass h, axial moment of inertia C, and transverse moment of inertia A, which is spinning on a rough plane with its axis vertical, n being the spin.

If $C^2n^2 < 4Amgh$, shew that a possible motion in the neighbourhood of the vertical is one in which the projection of the centre of mass of the top onto the horizontal plane describes approximately the equiangular spiral

$$r = r_0 \exp \frac{\sqrt{4Amgh-C^2n^2}}{Cn}\,\theta.$$

6. A uniform solid right circular cone is of mass M, height h, and semi-vertical angle α. The cone is rotating freely about its axis of figure when a point on the rim of the base is suddenly fixed. Prove that the axis about which it then proceeds to rotate makes an angle β with the axis of the cone where $\tan\beta = 5\tan\alpha/(2+3\tan^2\alpha)$.

7. A sphere of radius a and mass M is loaded so that, while its centre of mass remains coincident with its centre of figure, its principal moments of inertia are A, A, C. Shew that a steady motion is possible in which the sphere rolls on a rough horizontal plane with the axis of C inclined at a constant angle α and describing a cone with constant precessional velocity ω about a vertical axis at a distance c from the centre of the sphere. Find the component

of the angular velocity of the sphere along the axis of C for such a steady motion.

8. One end of the axis of a symmetrical top is fixed and the other slides freely along a light guide in the form of a vertical circle with centre at the fixed end of the axis and movable about a vertical diameter, which is fixed. Shew that if the guide is constrained to rotate about its fixed diameter with constant angular velocity ω, then (K = constant)

$$A\dot{\theta}^2 = K + 2(Cn\omega - Mgh)\cos\theta - A\omega^2\cos^2\theta.$$

Shew that it is possible to choose the initial conditions in such a way that θ decreases steadily from its initial value α (> 0) and tends to zero as $t \to \infty$. Find θ as function of t in that case.

9. A top of mass M, height of centre of mass h, axial moment of inertia C, and transverse moment of inertia A on a rough horizontal plane has its axis inclined at θ_0 ($\neq 0$) to the upward vertical. It is given a spin n about its axis and $C^2n^2 \gg 4AMgh$. Shew that the centre of mass drops a distance $2Mgh^2A\sin^2\theta_0/C^2n^2$ approximately, and that a precession with angular speed $(Mgh/Cn)[1-\cos(Cnt/A)]$ takes place.

10. A circular disc of radius a spins on a smooth table about a vertical diameter. Find the condition that this motion is stable.

11. A thin disc of radius a and mass m is spinning with angular velocity n about the normal through its centre which is fixed. Shew that if it is slightly disturbed, the axis precesses with frequency n.

If the motion is disturbed by a finite impulse of moment J about a diameter, find the maximum angle of deviation of the axis.

12. A thin circular disc of radius a rolls on a rough horizontal table. If θ and ψ are the angular coordinates of its axis, referred to the vertical and a fixed vertical plane, and if n is its spin about its axis, find the equations of motion for θ, ψ, and n.

Find the conditions of stability, if the disc is set spinning with angular velocity ω about a vertical diameter.

13. A uniform circular disc can rotate freely about its centre of mass which is fixed. Shew that the instantaneous axis of rotation describes, relative to the disc, a circular cone in simple harmonic motion. If the semi-angle

of this cone be α, shew that the instantaneous axis describes in space a circular cone of semi-angle $\alpha - \arctan(\frac{1}{2}\tan\alpha)$.

14. A hoop, of radius R, is rolling with its plane vertical along a straight line on a perfectly rough horizontal table. The velocity of the hoop is V. It is then slightly disturbed. Shew that if $V^2 > \frac{1}{4}gR$, then in the subsequent motion the plane of the hoop will oscillate about the vertical with frequency $(2\pi R)^{-1}\sqrt{1/3(8V^2 - 2gR)}$.

15. A uniform solid cube of side $2a$ is spinning with angular velocity ω around a diagonal which is vertical. Suddenly one of the edges, which does not meet this diagonal, is held fixed, and the cube subsequently rotates freely about this new axis. Find the condition that it will make complete revolutions about the new axis.

16. A top which is spinning on a perfectly rough table has principal moments of inertia A, A, C at its vertex. Prove that if the spin n of the top satisfies $C^2n^2 > 4Amgh\cos\alpha$ then there are two steady precessional motions of the top about the vertical with the axis of the top inclined at a given angle α to the vertical.

If the top is in one of these steady precessional motions and is then slightly disturbed, prove that the inclination θ of the axis performs a simple harmonic motion about $\theta = \alpha$, and find the period of the motion.

17. A particle is projected in latitude λ North with velocity V at an elevation α in a direction θ East of South. Prove that the time of flight is given approximately by

$$\frac{2V\sin\alpha}{g}\left[1 + \frac{2\omega V}{g}\cos\alpha\cos\lambda\sin\theta\right],$$

where ω is the angular velocity of the earth.

Prove also that the particle falls approximately

$$4\omega\sin^2\alpha[\cos\alpha\sin\lambda + \tfrac{1}{3}\sin\alpha\cos\lambda\cos\theta]\frac{V^3}{g^2}$$

to the right of the vertical plane of projection. Explain the approximations you make, and give the orders of magnitude of the differences between the true results and the approximate results given above.

18. If the bob of the Foucault pendulum (see § 4.3) is released from rela-

tive rest at the point $(a, 0)$, find the distance of closest approach of the bob to its equilibrium position $(0, 0)$.

19. A particle of mass m is free to move on a smooth horizontal table fixed on the earth at apparent latitude λ. It is moving in a potential $U = \frac{1}{2}mp^2r^2$, where r is the distance from O. At $t = 0$, the particle is projected from O with velocity U. Shew that with a suitable choice of polar coordinates the equation to the path of the particle is

$$pr = U \sin (p\theta/\omega \sin \lambda),$$

neglecting squares of ω/p, where ω is the angular velocity of the earth.

20. Discuss the motion of Foucault's rotor, which is a solid cylinder rotating around its axis suspended by an axis through its centre of gravity, the axis being in the East-West direction.

21. A gyroscopic compass is a gyroscope spinning about its axis, which is free to turn in a horizontal plane. Assuming that the centre of the gyroscope is at rest relative to the rotating earth, shew that the axis of the gyroscope can remain at rest relative to the earth, if pointing North.

The spin of the gyroscope about its axis is n, the angular velocity of the earth is ω, and $n \gg \omega$. Shew that, if the axis is slightly disturbed it will oscillate about the true North with a period which is approximately equal to $2\pi\sqrt{A/Cn\omega \cos \lambda}$, where A and C are the transverse and axial moments of inertia of the gyroscope, and λ is the latitude.

CHAPTER 5

1. The Hamiltonian of a system of two degrees of freedom is given by the equation

$$H = \tfrac{1}{2}(p_1^2q_1^4 + p_2^2q_1^2 - 2\alpha q_1),$$

where α is a constant. Prove that

$$q_1 = A \cos q_2 + B \sin q_2 + C,$$

where A, B, and C are constants.

2. Shew that the transformation

$$Q = \ln \left(\frac{1}{q} \sin p\right), \; P = q \cot p,$$

is a canonical transformation and find the generating function $S(p, Q)$.

3. Prove that the transformation

$$p = K\sqrt{2\alpha}\cos\beta, \qquad q = K^{-1}\sqrt{2\alpha}\sin\beta$$

is a contact transformation; find the generating function $S(\alpha, q)$, and apply this transformation to the problem of a one-dimensional harmonic oscillator.

4. Shew that the transformation

$$Q_1 = q_1^2+p_1^2, \qquad Q_2 = \tfrac{1}{2}(q_1^2+q_2^2+p_1^2+p_2^2),$$
$$P_1 = \tfrac{1}{2}\arctan(q_2/p_2)-\tfrac{1}{2}\arctan(q_1/p_1), \qquad P_2 = -\arctan(q_2/p_2)$$

is a contact transformation.

Apply this transformation to solve the equations of motion for the system whose Hamiltonian is $H = \frac{1}{2}(p_1^2+p_2^2+q_1^2+q_2^2)$, and compare this solution with the solution of the equations of motion in terms of the original variables.

5. Prove that the transformation

$$p_1 = \sqrt{k_1\alpha_1}\sin\beta_1+\sqrt{k_2\alpha_2}\sin\beta_2, \quad p_2 = -\sqrt{k_1\alpha_1}\sin\beta_1+\sqrt{k_2\alpha_2}\sin\beta_2,$$
$$q_1 = \sqrt{\alpha_1/k_1}\cos\beta_1+\sqrt{\alpha_2/k_2}\cos\beta_2, \; q_2 = -\sqrt{\alpha_1/k_1}\cos\beta_1+\sqrt{\beta_2/k_2}\cos\beta_2$$

is a contact transformation. Use this transformation to solve the equations of motion for the system with Hamiltonian

$$H = \tfrac{1}{2}p_1^2+\tfrac{1}{2}p_2^2+\tfrac{1}{4}k_1^2(q_1-q_2)^2+\tfrac{1}{4}k_2^2(q_1+q_2)^2.$$

Discuss the physical meaning of this Hamiltonian.

6. Shew that the transformation

$$x_i = P^2X_i+WP_i, \qquad p_i = P_i/P^2 \qquad (i = 1, \ldots, n),$$

with

$$P^2 = \sum_{i=1}^{n} P_i^2 \quad \text{and} \quad W = -2\sum_{i=1}^{n} X_iP_i$$

is a canonical transformation.

If the Hamiltonian of a system is given by the equation

$$H = \tfrac{1}{2}\sum_{i=1}^{n} p_i^2-\mu\Big[\sum_i x_i^2\Big]^{-\frac{1}{2}},$$

shew that in any solution for which the total energy is zero, the X_i remain constant in time and the P_i are linear functions of u with $u = \int P^{-2}dt$.

Comment on the results obtained in this problem.

7. A system of n degrees of freedom has a Hamiltonian given by the equation

$$H = \tfrac{1}{2}\sum_{r=1}^{n} p_r^2 + \sum_{r=1}^{n+1} (q_r - q_{r-1})^2 \quad (q_0 = q_{n+1} \equiv 0).$$

Use the contact transformation generated by the generating function

$$W = \sum_{r=1}^{n} \tfrac{1}{4} w_r^2 \left[\sum_{s=1}^{n} a_{sr} q_s\right]^2 \cot Q_r$$

with

$$a_{sr} = \frac{2}{\sqrt{(n+1)w_r}} \sin \frac{sr\pi}{n+1}, \quad w_r = 2 \sin \frac{r\pi}{2(n+1)}$$

to solve the equations of motion of the q_r and to find the normal coordinates of the problem. Comment on your results.

CHAPTER 6

1. Use the Hamilton-Jacobi equation to obtain an equation for the orbit of a particle moving in a two-dimensional potential $U = \mu/r$, describing the motion in terms of the coordinates $u = r + x$, $v = r - x$.

2. Use the Hamilton-Jacobi equation to obtain an equation for the orbit of a particle moving in a two-dimensional potential $U = \frac{1}{2}\alpha r^2$, describing the motion in terms of the coordinates u and v defined by $x = \cosh u \cos v$, $y = \sinh u \sin v$.

3. Use the Hamilton-Jacobi equation and elliptical coordinates to describe the motion of a charged particle in the field of two charges which are fixed at a finite distance from one another.

4. Use the Hamilton-Jacobi equation to shew that the orbit of a particle with Hamiltonian

$$H = \tfrac{1}{2}(p_1^2 + p_2^2)(q_1^2 + q_2^2)^{-1} + (q_1^2 + q_2^2)^{-1}$$

is a conic in the q_1, q_2-plane.

5. Use the Hamilton-Jacobi theory to discuss the Stark effect of the hydrogen atom, introducing parabolic coordinates. Expand the results obtained in powers of the electrical field strength and compare them with the results found in § 7.3.

6. For a certain system with two degrees of freedom the integral of energy is $H(q_1, q_2, p_1, p_2, t) = h$, another integral of motion is

$$F(q_1, q_2, p_1, p_2) = c.$$

Shew that there exists a function $\psi(q_1, q_2, h, c)$ such that

$$p_1 = \frac{\partial \psi}{\partial q_1}, \quad p_2 = \frac{\partial \psi}{\partial q_2},$$

and that the remaining integrals are

$$\frac{\partial \psi}{\partial c} = \text{constant}, \quad \frac{\partial \psi}{\partial h} - t = \text{constant}.$$

If

$$H = q_1 p_2 - q_2 p_1 + a p_1^2 - a p_2^2,$$

shew that there is an integral

$$p_1 q_1 + p_2 q_2 - 2a p_1 p_2 = \text{constant},$$

and complete the solution.

7. A particle of mass m moves with energy E in the x, y-plane in a potential U. If $w = \varphi + i\psi$ is an analytic function of the complex variable $z = x + iy$ such that

$$\left|\frac{dw}{dz}\right|^2 = 2m(E - U),$$

shew that a complete integral of the Hamilton-Jacobi equation is

$$S = \varphi \cos \alpha_1 + \psi \sin \alpha_1 + \alpha_2,$$

where α_1 and α_2 are constants of integration.

Determine the trajectories of particles projected from infinity with $E = 0$ into a potential

$$U = -\lambda^2 \mathrm{OP}^2/(\mathrm{AP}^2 \cdot \mathrm{BP}^2),$$

where A, B, O and P are the points $(a, 0)$, $(-a, 0)$, $(0, 0)$, and (x, y), and λ is a constant.

CHAPTER 7

1. Discuss the perturbation of a one-dimensional harmonic oscillator by

a linear or a quadratic term in the Hamiltonian and compare the result with the exact solutions of the equation of motion.

2. Discuss the perturbation of a one-dimensional harmonic oscillator by a quartic term.

3. Discuss the secular effects of crossed electrical and magnetic fields on the hydrogen atom.

CHAPTER 8

1. Discuss the longitudinal elastic vibrations of an infinitely long elastic rod by approximating this system through a discrete system of equal mass points connected by uniform massless springs. The forces are supposed to be purely harmonic, and the masses to be uniformly spaced. Consider the limit where the spacing goes to zero and obtain in that way the wave equation (8.101).

2. Shew that in the long-wavelength limit the propagation of longitudinal waves down an infinite line of equidistant atoms interacting only through nearest neighbour elastic interactions and with the masses of alternate atoms m and M corresponds to that down an equivalent homogeneous line.

3. A string of length $2L$ is stretched to tension T between two fixed points $x = -L$ and $x = L$. The density at the point x is $m/(2L-|x|)^2$. Find the equation to be satisfied by the frequency of small transverse vibrations.

4. A uniform string is stretched to tension T between $x = 0$ and $x = l$. A small variable transverse force of magnitude $F(x, t)$ per unit length is applied at time t at the point x. Find the transverse displacement as function of x and t.

A hammer gives a transverse velocity v to the part of the string $h-\delta < x < h+\delta$, and leaves the remainder of the string initially at rest. Find the subsequent displacement of the string and discuss the special case where h/l is a rational fraction.

5. One end of a uniform flexible chain of length l is attached to a vertical rod which rotates at constant angular velocity Ω. Neglect the effect of gravity so that the chain sweeps out a horizontal circle. Apply Hamilton's variational principle to derive the wave equation for small transverse vibrations, and find the frequency of the fundamental mode of vibration.

INDEX

GLOSSARY OF SYMBOLS

This glossary contains a list of the more important symbols occurring in the text. The page number indicates where these symbols are introduced.

a, force constant of simple harmonic oscillator 5
a, semi-major axis 16
$\mathbf{a}$, acceleration 2
A, principal moment of inertia . 80
$\mathbf{A}$, vector potential 46
B, principal moment of inertia. 80
$\mathbf{B}$, magnetic induction. 46
C, principal moment of inertia. 80
D_{kl}, inertial tensor 80
E, energy 4
E, Young's modulus 168
$\mathbf{E}$, electrical field strength . . . 46
$\mathbf{F}$, force 2
g, gravitational acceleration . . 33
G, gravitational constant . . . 17
G_l, functions occurring in kinematic relations 28
H, Hamiltonian. 96
$\mathscr{H}$, Hamiltonian density 173
I, action integral. 115
J_i, action variable 135
$\mathbf{J}$, angular momentum of rigid body 79
L, Lagrangian 35
$\mathscr{L}$, Lagrangian density. 172
m, mass 2
$M_{\odot}$, solar mass 88
M, angular momentum 7
$\mathscr{M}$, torque 79
$\mathbf{n}$, unit vector 109
p, parameter of ellipse 16
p, impact parameter 19
p, angular velocity around X-axis 77
p_k, generalised momentum . . . 37
p_0, negative energy 119
$\mathbf{p}$, linear momentum 3
$\mathbf{P}$, total linear momentum vector 108
q, angular velocity around Y-axis 77
q_k, generalised coordinate . . . 35
$q_k^{(0)}$, equilibrium value of generalised coordinate 50
$\dot{q}_k$, generalised velocity 35
q_0, time 119
Q_m, normal coordinate 53
Q_k, Fourier coordinate of $Q(x)$. 168
$Q(x)$, canonical variable for continous systems. 167
r, polar coordinate. 5
r, angular velocity around the Z-axis 77
r_k, independent variable ($=\dot{q}_k$). 95
R, Routh function 41
R, given by (6.144) 128
s, number of degrees of freedom 29
s, sound velocity. 175
S, generating function 102
S, Hamilton-Jacobi function. . 122
t, time coordinate 2
T, kinetic energy 3
$\mathscr{T}$, kinetic energy density . . . 171
u, eccentric anomaly 131
U, potential energy 4
$\mathscr{U}$, potential energy density . . . 171
v, transverse velocity 145
$\mathbf{v}$, velocity vector. 2
$\mathscr{V}$, Clausius' virial 23
w_i, angle variable 136
W, generating function 101
$\mathbf{x}$, position coordinate vector. . 2
$\mathbf{X}$, centre of mass coordinate vector 18
Z, atomic number 127

α_k , canonical variable 98
β_k , canonical variable 98
δ_{mn} , Kronecker symbol 56
ε , eccentricity 17
$\boldsymbol{\varepsilon}$, infinitesimal vector. 47
θ , polar angle 6
θ , Euler angle 76
κ , force constant in inverse square law 6
λ , perturbation parameter . . 147
λ_l , Lagrangian multiplier . . 31
λ_m , eigenvalue of the small-vibrations-matrix problem . . 54
μ , reduced mass 18
ν , frequency of simple harmonic oscillator 5
π , momentum density. 173
π_k , momentum in the continuous-system case. 170
ξ , displacement 168
ξ_k , Fourier component of the displacement 169
ρ , mass density 80
σ , $= 1/r$. 12
τ , period in Kepler orbit . . . 17
τ_i , periods of multiply periodic motion. 136
φ , polar angle 6
φ , Euler angle 76
ϕ , scalar electromagnetic potential 46
χ , true anomaly 130
ψ , Euler angle 76
$\boldsymbol{\omega}$, angular velocity vector . . . 79

$\cdot$, time derivative 3
$\boldsymbol{\nabla}$, gradient operator 4
[ ] , Lagrangian bracket 104
{} , Poisson bracket 104
$\delta/\delta\xi$, functional derivative 173